FRONTIERS OF SPACE

Herbert Friedman, General Editor

D1451492

ROCKETS INTO SPACE

FRANK H. WINTER

Harvard University Press · Cambridge, Massachusetts, and London, England · 1990

Library of Congress Cataloging-in-Publication Data
Winter, Frank H.
  Rockets into space / Frank H. Winter.
    p.   cm. — (Frontiers of space)
  Includes bibliographical references.
  ISBN 0-674-77660-7
  1. Rocketry—History. I. Title. II. Series.
TL781.W55   1990   89-24551
629.47'52'09—dc20   CIP

To the memory of Jules Verne,
who pointed the way

FOREWORD BY HERBERT FRIEDMAN

The evolution of modern rocketry from the visionary concepts of pioneers like Tsiolkovsky, Goddard, and Oberth to the U.S. Space Shuttle and the Soviet Energia rocket is one of the greatest technological miracles of our time. It has all happened so fast that the entire epoch encompasses barely one generation. The public of Goddard's time was completely unfamiliar with the basic principle of rocket propulsion, as shown by the following extract from a *New York Times* editorial of 1920: "That Professor Goddard, with his 'chair' in Clark College and the countenancing of the Smithsonian Institution does not *know* the relation of action to reaction and of the need to have something better than a vacuum against which to react—to say that would be absurd. Of course he only seems to lack the knowledge ladled out daily in high schools." At the end of World War II the United States had no rocketry more sophisticated than the bazooka, a small antitank rocket launched by infantrymen from a tube resting on the shoulder, and the JATO (or Jet-Assisted Takeoff rocket), which was mounted on an aircraft to increase its lift on takeoff. The Germans, however, had made greater progress: they shocked the Western world with their monster rocket, the V-2, capable of launching a 2,000-pound explosive payload from the European continent to the British Isles and other targets. An even more profound shock was the orbiting of the Soviet Sputnik in 1957. Western nations reacted immediately by accelerating rocket development efforts and by sharply increasing all scientific support.

The Apollo commitment—to send men to the moon and bring them back safely within the decade of the sixties—was, despite its great cost, strongly supported by the U.S. Congress and received with widespread approval in the free world. In the ensuing years American resources went into development of the Space Shuttle, while Soviet efforts produced a versatile stable of expendable rockets culminating in the giant Energia, capable of lifting more than 200,000 pounds into orbit.

Our century has been battered by the catastrophes of the Great Depression, World War II, the advent of the nuclear age, the Cold War, the Vietnam War, and now the increasing environmental peril. Although rockets, in the form of ballistic missiles, can potentially cause unprecedented destruction and horror, as vehicles into space they also offer an open-ended invitation to intellectual exploration and discovery, along with great opportunities for practical benefits for all mankind. It can be argued, as well, that ballistic missiles have deterred World War III for four decades.

In the current heat of controversy over national priorities in space and the need for a mixed fleet of manned and unmanned vehicles, Frank Winter's book provides a valuable perspective on the historical background of rocket development. It is dense with interesting detail, and filled with anecdotes and insights about the pioneers of rocketry and the various environments in which they pursued their technical goals.

Rockets into Space is the first volume in a new Harvard University Press series, *Frontiers of Space,* that will explore the various facets of space science and technology. Rockets are the essential underpinning of all space endeavors, and it is most appropriate to begin the series with a comprehensive history of rocket development.

CONTENTS

ACKNOWLEDGMENTS

This book was written comparatively recently, but much of it is based on research stretching back many years, during my tenure first as a historian and then as the curator of rocketry at the National Air and Space Museum. I must, therefore, be forever grateful to the late Mrs. Esther C. Goddard, widow of the American rocket pioneer Robert H. Goddard. She introduced me to the National Air and Space Museum and opened up a wealth of ideas and challenges to me, as I researched the achievements of Dr. Goddard as well as the history of rocket technology in general. In her continued encouragement of my research goals, Mrs. Goddard furnished me with much essential information. Since her passing, I have relied on the indispensable aid of Mrs. Dorothy Mosakowsky of the Robert H. Goddard Library, Clark University, Worcester, Massachusetts.

I also owe a very special debt of gratitude to Frederick C. Durant III, former Assistant Director of the Astronautics Department, National Air and Space Museum, who not only served as my first supervisor at the museum but who introduced me to many pioneers in the fields of rocketry and astronautics, including Arthur C. Clarke, G. Edward Pendray, and Frank Malina. Thanks to him, I was able to conduct invaluable interviews which furthered my education in the evolution of astronautics, and of rocket technology in particular.

Throughout the years, I have benefited from my friendship and professional association with rocketry and spaceflight authorities Mitchell R. Sharpe and Frederick I. Ordway III, who likewise helped with the preparation of this book. Another friend and rocketry pioneer, George S. James, offered many useful insights. Lee Saegesser, archivist of the

History Office, National Air and Space Museum, furnished me with extensive documentation for this project.

Dr. Martin Harwit, the present director of the National Air and Space Museum, deserves thanks for inviting me to participate in an interview with the late Hermann Oberth, one of the three great founders of astronautics. Professor Oberth's passing in late 1989, at the age of ninety-five, was a profound loss both to the world and to his friends. I am likewise grateful to Dr. Gregg Herken, chairman of the museum's Space History Department, for his continued encouragement of my research.

Dr. Milton Rosen, who served as development director of the Viking sounding rocket and subsequently became NASA's launch vehicle director during the critical years of the early 1960s, helped me in clarifying numerous technical and historical questions. George P. Sutton, another outstanding rocket pioneer, provided me with a firsthand account of the role of the V-2 in the development of large-scale rocketry in the United States. Dieter Huzel, whose career has spanned both Peenemünde and NASA, imparted valuable insights and data. The reminiscences of Darrell Romick were also extremely helpful to me as I composed the account of the early years of the Space Shuttle.

Finally, I owe my deepest appreciation to Dr. Herbert Friedman, who placed his faith in me to undertake this project; and to my wife, Fe Dulce, who, during a difficult two years, showed patience and understanding that were nothing short of remarkable.

ROCKETS INTO SPACE

INTRODUCTION

Spaceflight as a concept has permeated the world's literature—mythological, religious, astronomical, and other—since ancient Babylonia, but without the modern launch vehicle it would not have become a reality. This book surveys the history, current development, and possible future of space boosters. Though brief, it is designed to fill a significant gap, since most existing books on rocketry are general histories covering the entire field from combat weapons to boosters, or are studies focused upon specific vehicles. It also traces, where possible, key technological threads or ideas that led to the successful evolution of launch vehicles.

The reaction principle has in fact existed in nature for perhaps half a billion years. Mollusks such as squid, scallops, octopi, and cuttlefish move in this way by taking in water and forcing it out again, thus propelling themselves forward. Sir Isaac Newton was the first to define this principle—which also explains why a rocket flies—in his *Principia* of 1687. His classic Third Law of Motion says, "For every action there is an opposite and equal reaction." Implicit in this law is that rocket motion can work anywhere, even in the vacuum of space. The Greeks had actually demonstrated the principle two thousand years earlier, but apparently the knowledge was lost or ignored. About 360 B.C., Archytas of Tarentum built a wooden pigeon suspended from a string. Steam or compressed air was ejected from portholes in its body, and the bird "flew" in a circle. In the first century A.D., Hero of Alexandria proposed what he called an Aeolipile, described in his *Pneumatica* as a hollow ball revolving on pivots whose arms were placed over a cauldron partly filled with water. When a fire was lighted under the cauldron, the resulting steam passed through the arms into the ball and out of two bent tubes on opposite sides of the ball. This caused the ball to revolve, much like a rotating lawn sprinkler. Whether such devices were intended to demonstrate a physical principle or to be used merely as toys, we do not know.

For the earliest known spaceflight legend, we must turn to the Babylonians. On clay tablets dating from 2350–2180 B.C. is described and depicted the Assyro-Babylonian epic poem of Etana. Etana flew on the back of an eagle far into the sky toward the dwelling place of Ishtar (the planet Venus) but kept going until the eagle was exhausted, or became frightened, and then suddenly descended back to Earth. Elsewhere, in China, the lunar goddess Lady Cheng-O is supposed to have flown to the moon by drinking a magic elixir, a story joyfully retold every year during the Moon Festival held on the fifteenth day of the eighth month. In Norse mythology, the god Mani is driven to the moon in a chariot. And the Greek tale of Icarus is widely known: Icarus aspired to reach the heavens by means of wings affixed to his body with wax, but he came too close to the sun, which melted the wax and caused him to fall into the sea.

The ancient Greeks also produced the first novels of lunar voyages, *Vera Historia (True History)* and *Icaro-Menippus,* both written about A.D. 160 by Lucian of Samosota. In the former, Lucian's travelers are whisked to the moon when their sailing vessel is carried aloft by a great wind. In the latter, the hero Menippus prepares for his trip, like Icarus, by fastening on bird wings. He flies toward the stars until Jupiter, angered by his presumption, orders that his wings be clipped and that he be conveyed back to Earth by the god Mercury.

Throughout the centuries, animal power was clearly the favorite literary device for achieving spaceflight. Witness the four red horses drawing a chariot to the moon in *Orlando Furioso,* by the sixteenth-century Italian poet Lodovico Ariosto, and the flock of trained geese pulling a chair for the moon-traveler in Bishop Francis Godwin's *Man in the Moon* (1638). Mechanical means also appeared occasionally, like the mammoth spring catapult in David Russen's *Iter Lunare; or Voyage to the Moon* (1703). Yet with one outstanding exception, the rocket was never considered for spaceflight in any of these early fictional accounts.

That exception is *Voyage dans la lune (Voyage in the Moon),* written in 1649 by Cyrano de Bergerac, the dashing French swordsman, playwright, and satirist. The rocket was one of several spaceflight schemes devised by Cyrano, who also wrote *Histoire comique des états et empires du soleil (Comical History of the States and Empires of the Sun)* in 1652. Cyrano's flying machine consisted of a box fastened above tiers of small firework rockets and lit by a central fuse, but the box never made it to the moon, since the fireworks burned out and the machine fell back to Earth. Though this was probably the first suggestion anywhere of a rocket-propelled spacecraft, Cyrano wrote not as a scientist but as a satirist, a fact that is evident in his other literary devices for reaching space—for example, vials of dew strapped to the space traveler (who would be lifted into the air as the dew rose with the morning sun); and an iron "flying chariot," which would be propelled by magnetic attraction. (Lodestones, or natural magnets, would be continually thrown from the chariot, which would then be attracted toward the magnets and thus move forward.)

Not until more than two centuries later did two other writers, Achille Eyraud and Jules Verne, allude to rockets and space. Eyraud's *Voyage à Venus (Voyage to Venus)*, published in 1865, describes a spaceship propelled by a "reaction motor," a kind of water rocket. Though Eyraud used the correct principle, if not the most efficient propellant, his book went largely unnoticed. Verne, in contrast, produced world-famous masterpieces qualifying him as the "father of modern science fiction." In his classic *De la terre à la lune (From the Earth to the Moon)* of 1865, the astronauts sit elegantly in the hollow of a padded, well-equipped artillery shell which is fired into space by a huge cannon. In the book's 1870 sequel, *Autour de la lune (Around the Moon)*, the projectile is thrown off course by a meteor and winds up orbiting the moon rather than landing. Firework rockets steer the projectile away from the moon to prevent a crash-landing, but eventually the would-be lunar voyagers descend safely in the Pacific Ocean and are rescued by an American ship.

Verne's space novels do contain technical flaws (notably his choice of a cannon as a launching means), but to the extent that technology would then allow, he carefully researched the scientific possibilities and was remarkably prophetic. For example, he accurately described weightlessness in outer space; and with the help of his cousin, astronomer Henri Garcet, he correctly determined escape velocity at 7 miles per second (about 25,000 miles per hour).

Verne's novels bequeathed far more than spaceflight ideas. His natu-

Figure 1. Cyrano de Bergerac carried to the moon by rockets. From Cyrano de Bergerac, *Les Oeuvres diverses de Monsieur de Bergerac*, 1710. Originally published as *Voyage dans la lune*, 1649.

ral prose style and attention to scientific detail made spaceflight *believable,* and he thus played a significant role in inspiring the major founders of modern spaceflight theory.

Had the rocket itself been more developed by the 1860s, perhaps Verne would have thought of it as the main means of space propulsion. Rockets seem to have originated in China during the Sung Dynasty (A.D. 960–1279), then quickly spread to Europe probably via established trade routes. The first rockets were crude affairs, gunpowder-filled bamboo or pasteboard tubes with little power and unpredictable trajectories. The newly discovered gunpowder offered more promise as a weapon in the form of cannons and other firearms, and as a consequence, little attention was devoted to the improvement of rocket technology until the nineteenth century. Up to that time, rockets were used for little more than fireworks or signals. In 1804, however, the Englishman William Congreve "rediscovered" the rocket and thereafter developed an entire weapons system based on it. Elite rocket troops were organized, and many other countries followed this example. Average war rockets of the day, which were cased in iron and packed with gunpowder, had ranges of about a mile and a half. Meanwhile, cannons and other artillery improved tremendously as a result of technical advances during the Industrial Revolution, so that by the time Jules Verne's second space book appeared (1870), the rocket was again largely obsolete.

There were, nevertheless, many proposals for rocket or reaction-propelled aircraft in the nineteenth century. But not until the late 1800s were such devices seriously considered as a means of penetrating space.

Why, throughout a millennium of history, was the rocket overlooked as a potential space vehicle? The answer was ignorance of the fundamental principle of rocket flight. Despite Newton's formulation of the Third Law of Motion (and despite a classroom demonstration model of a steam reaction-propelled car by Dutch professor Jacob s'Gravensande about 1720), there arose two main theories of rocket flight. In one, propounded in 1717 by Edmé Mariotte of France, the rocket moved because the flame pushed against the surrounding air. The other was formulated a year later by Mariotte's compatriot, John T. Desaguliers: "[If the] rocket is open at the bottom, the action of the flame downward is taken away, but the upward force remaining pushes the rocket." In essence, Desaguliers correctly deduced that exhaust gases produce a forward momentum, though this is only part of Newton's law of equal and opposite reaction. Mariotte's reasoning, in contrast, was totally erroneous. Simply put, he believed that since air is needed for rockets to "push against," they could never enter a vacuum (that is, space).

The "air-pushing" school was the most widely accepted theory of rocket motion, accepted even by artillery experts and pyrotechnicians well into the twentieth century. As such, it was a long-standing impasse to progress. Yet the odd ideas of Hermann Ganswindt of Germany marked a near turning point in this thinking. Around 1890, Ganswindt conceived a reaction-propelled spaceship. Unfortunately, the eccentric Ganswindt espoused two divergent theories as to how it would work— he believed in both the "air-pushing" principle and the reaction principle. He maintained that for a "useful reaction" to take place in space, a spacecraft would have to throw off solid bodies. The fuel for his spaceship consisted of heavy steel cartridges with dynamite charges. The cartridges were fed machine-gun fashion into a chamber and exploded individually, imparting kinetic energy to the chamber and then dropping away. Shock absorbers protected the travelers, who were contained in a cabin that was connected to the explosive motor by a shaft. Adjacent to this motor were the two cartridge fuel "tanks." When sufficient velocity was reached, the motor was turned off and the ship coasted the rest of the way. Ganswindt unfortunately lacked the ability and patience to work out details and solve technical problems, but his lectures did much to arouse early public enthusiasm for spaceflight. His concept also represents a late-nineteenth-century benchmark of public understanding, or rather misunderstanding, of reaction propulsion as applied to spaceflight. It thus gives us a far greater appreciation of the tremendous strides made by the true founders of modern spaceflight theory: Konstantin Tsiolkovsky, Robert Goddard, and Hermann Oberth.

THE FOUNDERS OF
SPACEFLIGHT THEORY

Konstantin Eduardovich Tsiolkovsky, largely self-taught and almost totally deaf, was the author of the world's first theoretical studies of liquid-propellant space vehicles. This extraordinary man was born September 5, 1857, in the city of Izhevsk, almost six hundred miles east of Moscow. His Polish-born father instilled in him a passion for science and invention, though the elder Tsiolkovsky's own efforts at invention did not succeed.

Konstantin's early years were happy ones. He was extremely fond of reading and had a vivid imagination, often dreaming of what the world might be like without gravity. His life changed at age nine, however. While sledding he caught a cold, which developed into scarlet fever and eventually resulted in near-deafness. He became withdrawn, and studied at home because he could not go to school. At fourteen he discovered a mathematics book in his father's library and soon mastered the subject, then turned to natural science. Konstantin's father recognized his son's aptitudes and sent him to Moscow at sixteen to further his studies, though the family could not afford a university education for him. Life was not easy in Moscow. Konstantin diligently listened to free lectures in chemistry, astronomy, and other sciences with the aid of a homemade ear trumpet. He spent most of his small allowance on materials for chemistry experiments and scientific models, leaving a few kopeks for a diet of brown bread.

During this period, Konstantin discovered spaceflight. He devoured Verne's space novels, but another great influence in his thinking on spaceflight was philosopher-librarian Nikolai Fyodorov, illegitimate son of Prince Pyotr Gagarin (one of whose collateral descendants was

Figure 2. Hermann Ganswindt's reaction-propelled spaceship, 1891.

Yuri Gagarin, the first man to be rocketed into space). Uppermost in Fyodorov's often bizarre philosophy was mankind's need for *Lebensraum*, that is, space necessary for material and spiritual development: the human species, he believed, had to seek other habitable planets. Space colonization likewise became one of Konstantin's ultimate goals. On a practical level, Fyodorov helped his young friend by giving him food and clothing; he also set a regimen of study, enabling Konstantin to pass qualifying examinations and obtain teaching posts.

During his stay in Moscow, from 1873 to 1876, Tsiolkovsky began refining his thoughts on spaceflight and how to achieve it. This period is not well documented, but we do know that at sixteen he was struck by the idea that centrifugal force could be a means to his end. The project he had in mind sounds very similar to the giant waterwheel in Edward Hale's story "The Brick Moon," serialized in the United States in 1869–1870 and considered an early space-travel classic because of its suggestion that a satellite and space station could be launched by a wheel. Tsiolkovsky initially thought that his centrifugal machine could have "great consequences," but next morning he was convinced it would never work. He realized that air friction would have prevented the projectile from ascending beyond the lower atmosphere and that the astronauts would have been crushed at an early stage by the acceleration of the spinning wheel. Young Konstantin also toyed with perpetual motion as a way to spaceflight, but dismissed it, too, as impractical.

Finally he began to gain an understanding of reactive motion. On March 28, 1883, he wrote in his diary, *Free Space:* "Let us assume a barrel . . . filled with a highly compressed gas. If we open one of its valves, the gas will flow out of the barrel in a continuous jet and the elasticity of the gas which repels its particles into space will also repel the barrel continuously." Yet, at this point, he did not equate the device with a rocket.

In 1892 Tsiolkovsky moved to the provincial town of Kaluga, about a hundred miles southwest of Moscow. Here he published, during the next few years, a number of science fiction stories which served as convenient vehicles for expanding his speculations on spaceflight. In his story "Dreams of the Earth and Sky" (1895), he envisioned artificial satellites, space stations and colonies, solar motors, manned asteroids, artificial gravity, gravity-assist travel in space, interstellar travel, and the possibility of extraterrestrial life. But no mechanism for leaving the Earth's surface was described: his literary device for reaching space consisted merely of the sudden, mysterious disappearance of Earth's gravity. In 1896, however, Tsiolkovsky drafted the first ten chapters of a novel entitled *Outside the Earth,* in which he first came to grips with the technological approach. Although the book would not be published in complete form until 1918, the draft contained a detailed description of a rocket-propelled spacecraft. The cigar-shaped manned rocket was sixty-five feet long and used unspecified liquid propellants that ignited on contact, producing a "continuous, uniform blast" through a series of "pipe nozzles." (Here, he anticipated the development of self-igniting liquid propellants, now termed "hypergolic," which do not require ignition systems; nonhypergolic propellants require igniters.)

After his discovery of the rocket as the solution he had long sought, Tsiolkovsky cast fantasy aside for calculation and in 1896 began his groundbreaking scientific work, *Exploration of Cosmic Space by Means of Reaction Devices*. Part one appeared in 1903 in the fifth issue of *Nauchnoye Obozreniye (Scientific Review)*, under the title "A Rocket into Cosmic Space." The introduction set out his reasons for choosing the rocket. In addition to being able to operate in a vacuum (according to Newton's Third Law), the rocket has the following advantages:

1. it is cheaper and easier to make than a gigantic gun;
2. its combustion force is more uniform and constant than a gun's, and its acceleration can be changed, so that one can safely send men as well as scientific instruments into space;
3. its velocity can be altered in any direction, permitting navigation and soft landing on another planet;
4. during ascent, the rocket can be accelerated in the rarified layers of the atmosphere and in the space vacuum; when descending through the atmosphere, it can be slowed to prevent overheating.

Tsiolkovsky realized, furthermore, that precise control over rocket acceleration could be achieved only by using liquid fuels in conjunction with valves and pumps, not by using solids like gunpowder, which burned uncontrollably, all at once. Liquids were also denser and yielded more energy (calories) per pound than solids. Moreover, Tsiolkovsky proved mathematically that air "interferes" with the expansion of rocket exhaust gases and that air resistance slows rockets down; thus, rockets function more efficiently in space. He arrived at a fundamental formula for rocket motion—dated May 10, 1897, in the original manuscript—showing that a rocket's velocity is proportional to its exhaust velocity (the best propellants require a higher exhaust velocity). This formula also incorporated "specific impulse" (I_{sp}), now a standard measure of the energy produced by rocket engines and propellants. Specific impulse is expressed in seconds and is equal to thrust in pounds divided by fuel consumption in pounds per second. It is also equal to exhaust velocity divided by gravitational acceleration (32 feet per second per second). (Tsiolkovsky at that time favored a fuel composed of liquid oxygen and liquid hydrogen, which is theoretically ideal since it produces the highest exhaust velocity.) He demonstrated the importance of the rocket nozzle for directing rather than scattering rocket exhaust. And he calculated basic formulas for thrust, flight velocities as functions of fuel consumptions, rocket efficiency during ascent, the influence of gravity in a vertical ascent, the effects of air drag on rocket motion, and the effects of rocket acceleration on a passenger of average weight.

In 1903 Tsiolkovsky developed his first space-rocket design, though the drawing was not published until 1911. He chose a teardrop shape to reduce air drag and used this configuration for nine later models. The subsequent models, however, did not use liquid oxygen and liquid hydrogen, because the latter was then difficult to obtain and handle. The 1903 model had a straight, gradually flaring exhaust nozzle and incorporated what later became known as "regenerative cooling." In this technique, which eluded so many later researchers, one or more of the propellants circulated around the combustion chamber, thus cooling it. The propellant was also preheated, so that it burned more efficiently when finally injected into the combustion chamber. The rocket's direction was determined by gyroscopic controls and by movable vanes in the path of the exhaust, anticipating the graphite exhaust vanes of the V-2 rocket developed in the 1940s. Passengers, equipment, oxygen dispensers, and air purifiers were placed in the rocket's nose.

Tsiolkovsky's ideas evolved constantly. In *Investigation of Universal Space by Means of Reactive Devices* (1911), he confirmed escape velocity at 25,050 miles per hour, vividly described gravitational forces during takeoff and sudden zero-gravity, proposed a method for steering a rocket into Earth orbit (orbit insertion) by means of an additional rocket thrust, and correctly computed orbital velocity (17,900 miles per hour) for satellite launches. Tsiolkovsky suggested that launches take place on Earth's equatorial plane, to take advantage of the planet's rotation in order to reduce escape velocity and hence propellant load. (About sixty years later, the European Space Agency chose French Guiana as their launch site for this very reason.) He computed flight time to the moon (4.8 days, a figure borne out by the Apollo missions), worked out plans for establishing rockets in lunar orbits, devised soft lunar touchdowns, and foresaw that one would need larger rockets to reach heavy planets like Jupiter and Saturn and smaller rockets to reach lighter ones. Yet Tsiolkovsky's fertile mind would not rest. He speculated at length on "life-support" systems, proposing that a rocket could regenerate its oxygen supply by incorporating a greenhouse, which would take in carbon dioxide and expel oxygen. He envisioned more powerful "reactive devices" driven by atomic energy (radium) and electromagnetic systems, and projected a series of orbiting manned space stations that could collect solar energy or serve as bases for colonizing the cosmos.

In later works, Tsiolkovsky investigated the energy potential of radium, light (photon propulsion), and various chemical fuels (kerosene, alcohol, and methane—all in combination with liquid oxygen, abbreviated today as "lox"). He analyzed lightweight, heat-resistant pumps (1926); conceived Space Shuttle–like rocket airplanes that made powerless glide landings (1924–1926); drew up a plan for static-testing lox-hydrogen (or hydrocarbon) rocket engines (1927); and devised the important concept of multistage space rockets, or what he called "cosmic trains" (1926–1929), an idea he had already fictionalized in his 1920 story "Outside Earth." A multistage or step rocket consists of separate stages used in succession, each stage being jettisoned after use; the top stage contains the payload. Such a vehicle can reach a much higher velocity than a single-stage rocket with the same initial weight, propellant capacity, and payload weight.

Historically, step rockets are very old, however. They were used to make complicated fireworks, as described in manuscripts by Conrad Haas of Rumania (1529), Casimir Siemienowicz (1650), and others. In 1909 American rocket pioneer Robert Goddard thought of "multiple" rockets for upper atmospheric research, and in 1914 he took out a U.S. patent for such a vehicle, though it used solid fuel. In 1911 Frenchman André Bing was granted a Belgian patent for a liquid-fuel step rocket. In 1923 the Rumanian Hermann Oberth published details of his two-stage Model B liquid-fuel rocket. Tsiolkovsky's versions, described in his book *Cosmic Rocket Trains* (1929), differed markedly from other multistage vehicles in that the stages were joined horizontally, like real trains. They moved along an inclined runway, then took to the air, headed into space, and dropped away one by one until the final stage reached escape velocity. The step principle for spaceflight was vindicated (in modified form) twenty-eight years later with the launch of Sputnik 1, at which point the Soviets began hailing Tsiolkovsky posthumously as the "father of cosmonautics."

But for all his monumental theoretical work, Tsiolkovsky's real influence on spaceflight development is hard to assess. Many of his books were published at his own expense and not widely circulated, even in his native country. Furthermore, he remained in provincial Kaluga and did not participate in organized USSR rocketry activities, which began in the 1920s. It is true that his theoretical genius was recognized by the new Soviet state: in 1919 he was elected to the Socialist Academy (later the USSR Academy of Sciences); in 1921 he was granted a pension for life; and when he died in 1935, the Soviet papers afforded him glowing tributes. In the 1920s and 1930s the writers N. A. Rynin and Yakov Perelman, who made spaceflight a popular topic, publicized him widely, and his name, if not the details of his work, became generally known in the West from this period. Yet none of his works was translated into a Western language until the late 1940s. More important, in the West, Robert Goddard and Hermann Oberth had independently arrived at similar theories, long before Tsiolkovsky's fame had spread beyond Kaluga.

Like Tsiolkovsky, Robert H. Goddard was shy, preferring to labor alone on his spaceflight ideas. But whereas Tsiolkovsky was a theorist, Goddard was an experimenter, though initially he did undergo an almost feverish theoretical phase.

Goddard was born October 5, 1882, in Worcester, Massachusetts. His father was a bookkeeper and, like Tsiolkovsky's father, a part-time inventor. There were other parallels. Goddard was a great admirer of Jules Verne. H. G. Wells's tale of a Martian invasion, *The War of the Worlds* (1898), made a deep impression—so much so, that on one unforgettable day, when he had climbed a backyard cherry tree to trim some branches, he lapsed into a daydream in which he voyaged to Mars. Like Tsiolkovsky's first hypothetical space vehicle, Goddard's whimsical craft operated by means of centrifugal force. Both pioneers were sixteen when they developed their first important ideas. And when young Goddard contracted tuberculosis, it affected him as scarlet fever had affected Tsiolkovsky: he became more reclusive, a voracious reader of science books, and a dreamer, often imagining that he had conquered gravity.

There the similarities ended. Goddard missed many schooldays because of his illness, but nevertheless went on to acquire a first-rate academic education. He received a bachelor's degree in science from

Worcester Polytechnic Institute in 1908, a master's degree from Clark University, Worcester, in 1910, and a doctorate in physics from the same school in 1911. Eventually, in 1919, he became a full professor at Clark.

Goddard's diary reveals that as a young man he delighted in setting off skyrockets on the Fourth of July; but he, like Tsiolkovsky, worked long and hard before making his own discovery of the rocket as *the* instrument of spaceflight. The daydream that had started in the cherry tree had so affected Goddard that he noted the date, October 19, 1899, in his diary and thereafter always quietly celebrated "Anniversary Day." He quickly abandoned centrifugal force as a viable approach to the launch problem, and experimented with wooden models of flying contraptions, none of which worked. At the same time, he began reading Cassell and Company's *Popular Educator,* a gift from his father. He encountered Newton's Third Law of Motion, but could not immediately grasp its connection with rocketry and spaceflight.

From 1902 to 1904, Goddard spent much time exploring the launch capabilities of a machine-gun device that fired bullets downward. He sought to take advantage of recoil according to Newton's principle, but his physics remained faulty and this invention, too, led nowhere. The next year he devised an explosive mechanism that used gunpowder, but it had an air intake and did not seem suitable for space. (He did not at that point know that gunpowder and other combustibles can actually burn in the vacuum of space, because they carry their own oxygen atoms for combustion.) The more his education progressed, the more spaceflight ideas flooded into his brain. After graduation from high school, he bought some notebooks and began systematically recording his evolving concepts.

Goddard's new modus operandi was to brainstorm in his notebooks, proposing all sorts of possibilities, and then to play devil's advocate, questioning, rejecting, introducing new ideas, criticizing again, returning to older notions. Among his earliest entries, dated January 1906, is the description of a magnetic reaction using electromagnetic generators. He considered various forms of this idea but always seemed to reach the same conclusion: that such a system was too costly in terms of energy expenditure and loss. Other 1906 entries concerned Hertz (radio) waves, acoustic waves, solar energy, and other radiation sources which, Goddard believed, might be used as long-distance reactive means to move masses. "Is it possible to transmit impulses through the ether [space] by which . . . bodies can be moved?'' he asked. But the Hertz principle was probably just radiation pressure, he reasoned, and would not provide sufficient energy. The same limitation applied to solar energy and acoustic waves.

Early in 1906 he devoted some thought to atomic energy, but reasoned that radioactivity would generate too much heat to be a practical source of power. For years thereafter, Goddard doubted that atomic energy could ever be controlled, yet he well appreciated its potential and never completely gave up speculating about its feasibility, or that of solar energy, at least insofar as they might be used for a launch in space (that is, from a space station).

Levitation was not omitted from his list of possibilities. Neither were "electric and other guns, . . . artificially stimulated radioactivity, artificial atoms of great energy consisting of moving positive and negative charges, . . . repulsion of charged particles, reaction against displacement currents in space, repulsion of highly heated material particles at the focus of parabolic mirrors, . . . and several other ideas.'' (On September 6, 1906, Goddard suggested ion propulsion, which led to a 1915 patent and the earliest known ion propulsion experiments, carried out in the 1920s, in part by some of his students.) Even Earth's gravitational field was considered, but it seemed to him "highly probable that the electric field of the Earth cannot possibly be used to expel a body from the surface of the Earth.''

Interspersed with these proposals were discussions of space hazards like meteor swarms, which could damage the hull of the space "car.'' Goddard speculated that instruments might be able to detect meteors by bouncing waves off them, and that the ship could be protected with "extremely hard armor,'' perhaps made of tantalum. Other ideas involved launching an explosively propelled apparatus into space from a balloon (June 1907); using a solar energy boiler on board the spacecraft (January 1908); sending a camera to photograph other planets and bringing it back to Earth (June 1908); orbiting a planet so as to decrease landing speed (June 1908); sending automatic signals from other planets (August 1908); steering spacecraft automatically by means of photosensitive cells (October 1908); and sending flares to the dark side of the new moon in order to observe hitherto unseen lunar features (December 1908).

The main problem was propulsion, and this Goddard summed up as having two essential facets: finding an energy source and devising a means of reaction "against the ether." By January 1908 he had returned to explosives, cordite in particular. (Cordite was a "double-base" or "smokeless" propellant, consisting of nitrocellulose and nitroglycerine and yielding more energy than gunpowder.) He now knew that explosives could work in a vacuum. But an explosion took an instant. The "car" would be propelled a short distance only. It required "replenishment." Thus, "in leaving the Earth, packets of explosives might be shot up at the car, a rocket action most intense while the packet was passing through dense air." This process could be repeated, so that eventually the craft would be "far enough away to go by itself."

Goddard seemed on the verge of "discovering" the rocket as the most practical solution to the launch problem. In fact, it took another year to achieve the breakthrough. His constant questioning brought him full circle: back to electrostatic energy and the electric gun. On January 24, 1909, when he finally did think of the space rocket, the fuel he envisioned was not weak gunpowder but double-base powder, in a multistage vehicle. This solution he called "slow propulsion by explosives." But characteristically, Goddard was still not convinced. What bothered him most was the rocket's inefficiency. On February 2, 1909, he observed: "With ordinary [powder] rocket mixtures, solid particles are ejected which burn, causing friction and loss of heat . . . Try, if possible an arrangement of H and O [hydrogen and oxygen] explosive jets, with compressed gas in small tanks which are subsequently shut off—giving perhaps 40 or 50 percent [efficiency]. To get even 50 percent efficiency, it will probably be necessary to have small explosive chambers and jets, into which the explosive (not too violent) is fed." Here was Goddard's first tentative suggestion for a controllable, efficient liquid-propellant hydrogen and oxygen space rocket. All the while, he was unaware that an obscure, deaf Russian schoolteacher had already reached this conclusion.

In his step-by-step approach, Goddard began refining the liquid-fuel idea. By June 1909, he had thought of other liquid combinations, since he knew that lox and hydrogen could be produced only by ultra-cold compression and that the "necessary refrigeration plant" was extremely expensive. He suggested a mixture of nitrogen tetroxide and ethane, and proposed cooling the combustion chamber by means of "the expanding liquid gases," or regenerative cooling. Two days later, he worked out a liquid-fuel system in which the propellants were

Figure 3. The founding fathers of spaceflight. Clockwise from upper left: Konstantin Tsiolkovsky, Robert Goddard, and Hermann Oberth.

forced into a combustion chamber under pressure. By August 1910, he had developed a general mathematical theory of the ideal liquid oxygen–hydrogen rocket.

All these remained paper concepts until February 9, 1909, when Goddard performed his first experiment. Like Tsiolkovsky, Goddard learned that a propellant's exhaust velocity was the key to determining rocket performance. Using the physics laboratory of the Worcester Polytechnic Institute, he attempted to measure the ejection velocity of gunpowder mixtures ("deflagrating substances") and found them to be roughly a "few hundred feet per second." Obviously more experiments were needed, but these were precluded by lack of funds and time. In early 1915, his schooling behind him, Goddard resumed his experiments, and in 1917 was aided by a research grant from the Smithsonian Institution. The culmination of the solid-fuel experimentation was his famous Smithsonian report, published in 1919 as *A Method of Reaching Extreme Altitudes*. This publication established Goddard as the preeminent researcher in the field of rocketry. The following year he began his work on liquid fuels, to which he dedicated the rest of his life.

In the winter of 1905, when Goddard began cramming his notebooks with spaceflight schemes and Tsiolkovsky was refining his first liquid-fuel rocket design, eleven-year-old Hermann Oberth read Verne's *From the Earth to the Moon*.

An unusually gifted child, Hermann questioned many technical details in the story and used the laws of falling bodies to calculate the escape velocity of Verne's space vehicle. Verne's figures were correct, but Hermann was sure that a giant cannon was not the answer to spaceflight. Terrific air resistance would have crushed the manned projectile before it ever left the cannon barrel. Most puzzling of all was Verne's use of rockets to retard the fall of the projectile on the moon. At first, Hermann thought that air would have been needed for the gases to push against. Then he reasoned further and came to an intuitive understanding of the principle of reactive motion ("When someone jumps from a boat to the shore, the boat will receive an impulse in the opposite direction"). Hermann reread Verne's novel until he had memorized it. At age thirteen, he computed the force of gravity on Verne's space travelers. "They would be flattened into pancakes," he concluded. "A cannon is not good for spaceflight. It must be done with a rocket." Sixteen years later, in 1923, Oberth wrote a book on this solution which probably exerted more influence on the development of spaceflight than the work of either Tsiolkovsky or Goddard. His book was, in fact, the cornerstone of the Space Age.

Oberth was born June 25, 1894, in Sibui, Transylvania, which was then part of Austria-Hungary and which in 1918 became Rumanian. But he was of German heritage and afterward became a German citizen. His father was a physician, a fact that accounted for Hermann's fascination with the behavior of the human body in spaceflight. One time while swimming, for instance, he discovered some of the important physical effects of weightlessness. And by subjecting himself to physiological experiments, he concluded that human beings could briefly withstand the forces of rapid acceleration. He initially studied medicine at the University of Munich, then switched to physics and astronomy. During World War I he served in the Austro-Hungarian infantry on the Russian front, where he was wounded and later transferred to a medical unit as an ambulance sergeant. Following the war, he resumed the study of physics at the University of Cluj (Klausenburg) in Romania, and at the universities of Munich, Göttingen, and Heidelberg in Germany, but was not awarded a doctorate: his dissertation on spaceflight was rejected by Max Wolf, an astronomer, since, recalled Oberth, "it dealt mainly with physical-medical subjects." Oberth was awarded a certificate, however, and became a high school teacher. The University of Cluj also bestowed upon him the title of Professor for his 1923 book *Die Rakete zu den Planetenräumen (The Rocket into Planetary Space)*, which grew out of his dissertation.

In 1909, at the age of fifteen, Oberth had designed his first manned, multistage rocket. It used a solid propellant (moistened guncotton) and had crude nozzles with regulating valves for adjusting the pressure. Oberth then conferred with gun and gunpowder experts and, like Tsiolkovsky and Goddard, concluded that powder-fueled rockets were too weak and uncontrollable for interplanetary flight. He turned to liquid propellants and in 1912 drafted his first plan for a vehicle fueled by liquid oxygen and liquid hydrogen. In the fall of 1917, in a report to the German War Department, he proposed a gyroscopically controlled long-range liquid-fuel rocket. It was very similar to the V-2 rocket of World War II but was larger and less complicated. The missile stood 82 feet tall, was 16.5 feet in diameter, weighed 10 tons, and burned liquid air and watered alcohol. The War Department was unimpressed and turned it down. By contrast, Oberth's book was soon attracting an enormous amount of attention.

Die Rakete zu den Planetenräumen was eighty-seven pages long, yet covered the entire spectrum of manned and unmanned rocket flight: liquid-fuel rocket construction, propulsion, inertial guidance and navigation, aerodynamics, thermodynamics, flight mechanics, pretesting, life-support systems, spaceflight hazards and their remedies, bioastronautics (including the psychological effects of spaceflight), reentry and recovery techniques, telescopic tracking, and the various applications of rocket technology to spaceflight. In his introduction, Oberth set out to "prove" four premises: (1) that building rockets capable of reaching the upper atmosphere was within the bounds of then-current technology; (2) that with "additional refinement," rockets capable of penetrating outer space could be constructed; (3) that with modifications, the same rockets could safely carry men; and (4) that the use of such rockets could have profitable applications. Oberth emphasized that he was presenting not blueprints but hypothetical standard models, and that the values in some of his equations were only approximate. A few critics, however, ignored these warnings.

The first section of the book introduced fundamental principles such as thrust, and formulas for the most favorable velocity. Oberth also adapted aerodynamic standards for artillery projectiles, in order to derive an approximate shape of a prototypical rocket that could move at sonic and supersonic speeds in air and space. The hypothetical rocket was also multistaged. Oberth proposed liquid oxygen and liquid hydrogen for the upper stage (termed the "H.R.," or Hydrogen Rocket) and lox and an alcohol-water mixture for the lower stage (the "A.R.," or Alcohol Rocket). Fuels were injected in a fine vaporized form into hot streams of oxygen. Here, Oberth presented the earliest known description of rocket propellant injectors ("diffusors"). The basic nozzle was funnel-shaped, a type called a convergent-divergent De Laval nozzle, named after Swedish engineer Carl Gustaf De Laval, who invented it in the 1880s to deliver steam at maximum pressure to turbine blades. In fact, Oberth ingeniously used turbine discharge theory to arrive at equations for optimum rocket nozzle design. (The De Laval nozzle, which Tsiolkovsky and Oberth also advocated, became standard on all rockets.) Oberth likewise suggested regenerative cooling.

Die Rakete next dealt with rocket motion and air resistance. Related topics were launch angles and trajectories, including orbital paths around the Earth (that is, rocket stages as artificial satellites), and the advantage of launching toward the east (as Tsiolkovsky had similarly suggested in 1911), to take advantage of the direction of the Earth's motion so that less fuel would be needed. In addition, Oberth touched upon the effect of super-cold liquid hydrogen on metals and the way in which low-temperature fuels are handled, a science that is today called cryogenics. At the same time he recognized the need for thin, internally pressurized walls that would provide maximum lightness yet maintain their strength.

Part II of the book discussed constructional details for a "Model B" rocket, which today would be called a "sounding" rocket, since it "sounded" or explored the composition and temperature of the upper atmosphere and the fringes of space. (Model A is not mentioned but is presumed to be the lower, A.R. stage.) Model B was 16.5 feet long, had a maximum diameter of 22 inches, and weighed 1,200 pounds. Oberth gave weight breakdowns, including those for propellants, engine, pumps, and fins. For attaining a high initial acceleration, he suggested adding an alcohol-fueled booster rocket. This weighed 440 pounds loaded. (As alternate means for boosting the rocket, Oberth proposed a gun and a high-altitude balloon.) He recommended that aluminum alloy and other specific metals be used as construction materials, and that the vehicle be insulated. Based upon propellant exhaust velocities, an expected combustion time of 36 to 40 seconds, nozzle ratios, and other parameters, Oberth calculated a peak altitude of 1,220 miles.

There was lengthy discussion of pumps and valves, the igniter, the combustion chamber and nozzle (lined with fire-resistant material, such as asbestos or "fire clay"), and the electrically operated gyroscopic stabilizer. Oberth even detailed the rocket's instrumentation, such as the spring-operated acceleration indicator, manometers for recording internal pressure levels, barometers, thermographs, and air sample collectors. Stabilizing fins for the A.R. were folded down during ascent and opened on descent, serving as a parachute, though the H.R. used a real parachute. Oberth also devoted considerable space to the need for pretesting rocket components, mainly by means of water pressure and static firings.

The third and final part of *Die Rakete* concerned the modification of the Model B rocket, so that it could carry men into space. "One would first have to determine experimentally," Oberth began, "how much pressure a human could withstand without adverse effects." Here he summed up his years of observations and experiments on how much the human body could withstand during aquatic dives, weightlessness in swimming pools, and so on. He was particularly astute in noting that the body can bear more "g-loads" (gravity-loads) when prone, which is indeed how astronauts are placed in modern launch vehicles for lift-offs. He also discussed the psychological effects of "abnormal pressure conditions," possible accidents during rocket ascents, and dangers in outer space (stray meteors, and so on). Interestingly, Oberth favored ascents and descents from water, since he believed that astronaut recovery could be carried out more easily on water than on land.

Of the rocket itself, Oberth suggested that the hydrogen rocket (H.R.) need not burn continuously in space but that it could be stopped and restarted by the pilot. This anticipated modern restartable liquid hydrogen–lox engines, such as the Centaur. But more fascinating, particularly to the reader of 1923, were Oberth's mission plans and "prospects" for the manned rocket. Presaging a future argument justifying spaceflight, he cautioned that "the usefulness of a scientific discovery cannot be judged in advance."

To begin with, Oberth suggested that spacemen wear modified diving suits as spacesuits to perform physical or physiological experiments outside the orbiting ship. Telescopes "of any size could be used, for the stars would not flicker." Investigations of radiating energy "from various regions of the universe" might be carried out, and the rocket orbited around the moon to explore the unknown side. A "miniature moon" (an artificial satellite) and a large reflecting-mirror "observation station" (space station) might also circle Earth as aids in geographic research, exploration, ship navigation, military reconnaissance, spotting and melting icebergs, and weather observation and modification. They could also serve as a platform for space weapons. Finally, the station could function as a fueling and launch point for "visits to other celestial bodies."

Die Rakete contained endless complex equations, but also enough readable and provocative topics for laymen so that it became an instant success. Unwittingly, the technical critics added to its fame by engaging in often heated debates on the finer points of Oberth's theories. The possibility of spaceflight became a significant popular fad, not only in

Germany but in many other countries as well. Indeed, Oberth's 1923 book virtually started an international astronautical movement. Spaceflight articles appeared by the score; other spaceflight authors emerged; and rocket societies soon sprang up in several countries. In 1925 a second edition of Oberth's work appeared, and in 1929 it was published in considerably expanded form (more than 400 pages) under the title *Wege zur Raumschiffahrt (Ways to Spaceflight)*.

Wege now contributed a wealth of new ideas. Many were Oberth's own, but others were refinements of concepts from the now impressive body of spaceflight literature that had evolved since the publication of *Die Rakete* in 1923. Historically speaking, this makes it difficult to determine "firsts." In any case, among the more interesting ideas in the 1929 book were: an electromagnetic gun for launching manned space projectiles (similar to Goddard's idea), space telescope designs (Goddard had also worked out some configurations but never published them), geological lunar exploration, lunar mining, colonies on Mars, asteroid observatories, kerosene and benzine as fuels, deceleration through the use of rockets, multiple nozzles, a reconnaissance rocket with a movie camera, cell growth experiments in zero-gravity, solar-power generators to "send electricity to the Earth" (now called power satellites), solar corona studies in space, testing Einstein's theory of relativity (would light travel in a straight line in space?), a long-range military liquid-fuel rocket with poisonous gas warhead, and interstellar flight. Oberth also came close to suggesting computerized inertial guidance systems for long-range rockets and a method for ablative cooling. (The latter is a way of cooling a spacecraft on reentry into the Earth's atmosphere by using a certain type of material, today usually in conjunction with a plastic and fiberglass shield, on the spacecraft's nose; as air friction heats the nose, the material slowly melts, thereby carrying the heat away and protecting the interior of the vehicle.)

Wege zur Raumschiffahrt was also a tool that enabled Oberth to counterattack his critics and to level his own barrages against other spaceflight theorists. For example, he believed that his liquid-fuel rocket system, in which the exhaust gases flowed uniformly, was more efficient than Goddard's solid-fuel type, in which the gases flowed intermittently. Oberth also objected to the use of the "rocket airplane" (a craft resembling today's Space Shuttle, conceived by Austrian researcher Max Valier) because his calculations told him there would be excessive heating from air friction (the modern Shuttle's heat-resistant tiles solved this problem). He also believed that the airplane was "not suitable for vertical ascent," that its aerodynamics were wrong, and so on. He did note, however, the practical proposal made independently by Valier and Tsiolkovsky: that the airplane return to Earth by simply gliding down, just as the Space Shuttle does today.

This leads to critical and inevitable historical questions: Did Tsiolkovsky and Goddard influence Oberth? And what was the individual and collective impact of all three of them on the international spaceflight movement and the subsequent state of the art?

Oberth later claimed he first heard of Tsiolkovsky in 1924, when Tsiolkovsky sent him a copy of *Rakyeta v kosmeetcheskoye prostranstvo (The Rocket into Cosmic Space),* published in Kaluga a few months after *Die Rakete* appeared. Oberth had *Rakyeta* translated by one of his students, a Russian emigrant, although the book contained a preface in German by Aleksandr L. Tchiyevsky lamenting that Tsiolkovsky had already thought of many "foreign" spaceflight ideas (he meant Oberth's) twenty years before. "The above information is not meant to establish K. E. Tsiolkovsky's priority," Tchiyevsky went on, but "[shows] the almost criminal indifference of our countrymen toward intellectuals . . . Do we always have to get from foreigners what originated in our boundless homeland and died in loneliness from neglect?" The implication is that Tsiolkovsky's works were not wholly appreciated in his own country (at least before the 1920s), that they were unknown to the West until the publication of *Die Rakete,* and that they played no role in initiating the space travel movement outside Russia.

As for Goddard, Oberth wrote in a postscript to the 1923 edition of *Die Rakete* that only after this work had been typeset did he become aware of Goddard's report of 1919, *A Method of Reaching Extreme Altitudes.* Oberth also noted that Goddard "was able to experiment, . . . whereas I attempted mainly a theoretical treatment . . . For this reason, our works complement each other." Even so, Goddard reported on solid-fuel experiments and only very briefly mentioned the possibility of liquids (oxygen-hydrogen). Moreover, he did not discuss *manned* spaceflight and confined his speculations on ion rockets and other concepts to his notebooks. Tsiolkovsky and Goddard technically started earlier and could claim many priorities, but their earliest work remained unknown to the public. Thus, by virtue of his thoroughness and the fact that his ideas were openly published and sparked a widespread movement, Oberth alone deserves the title "Father of the Space Age." Moreover, his book *Die Rakete,* upon its publication in 1923, immediately found a number of "disciples"—researchers who supported his views, disseminated them, and built upon them in their own work.

One of Oberth's first champions was Max Valier, an ardent advocate of the pseudoscientific notion of "glacial cosmogony," which held, among other things, that the moon and planets are coated with ice. Valier was so intent on proving this bizarre theory that he sought to promote Oberth's work in all the journals available, so as to foster the development of interplanetary flight. He wrote to Oberth pleading for a collaboration. Oberth was grateful for any support, but he himself did not believe in glacial cosmogony. Although busy with his teaching, he did provide Valier with calculations which shortly appeared in Valier's *Der Vorstoss in den Weltenraum (Advance into Interplanetary Space),* published in 1924. One thus sees strange bedfellows in the early history of rocketry. The book contained glaring errors but was indeed enormously successful in popularizing Oberth's ideas, and it went through five editions. Valier afterward conducted sensationalistic rocket experiments with rocket cars, and died in 1930 while working on one of them.

In 1925 Walter Hohmann, a city architect from Essen, published *Die Erreichbarkeit der Himmelskörper (The Attainability of Celestial Bodies),* which, despite its almost mystical title, was a more technically significant and lasting contribution to the literature. Hohmann, who actually began making his calculations in 1914 as a hobby, focused upon theoretical problems of space navigation and fuel requirements for departing from Earth, orbiting planets, and landing on them. The famous Hohmann ellipses, in which spacecraft take advantage of the "gravity-assist" of planets by naturally flying into their orbits, has been used in U.S. Mariner and other programs.

Another early classic, Hermann Noordung's *Das Problem der Befahrung des Weltraums (The Problem of Spaceflight),* published in 1929, picked up on Oberth's suggestion of a space station: it was the first engineering study of space station construction, living, and application. Noordung was actually a pseudonym used by Captain Hermann

Potočnik of the Austrian Army, who died the year the book appeared. His work was so comprehensive that no other space station study appeared until the mid-1940s.

But not all the early pioneers were Germanic. Robert Esnault-Pelterie, who made great advances in early aviation as well, published *L'Exploration par fusées (Exploration by Rockets)* and *L'Astronautique (Astronautics),* in 1928 and 1930, respectively. REP, as he was also called, calculated rocket flight times to the planets as early as 1912. In later works, he was the first to scientifically analyze the significance of relativity theory for spaceflight, and to assess the possibilities of extraterrestrial life. He also investigated nuclear propulsion. More subtly, a new word entered the language: "astronautics" (the science of spaceflight), coined in 1928 when first applied to the annual International REP-Hirsch Astronautics Prize for the most outstanding scientific study of spaceflight. (André-Louis Hirsch was REP's banker friend who provided the prize money of 5,000 francs.) Yet by 1930 only the theoretical foundations of astronautics (or "cosmonautics," as the Soviets called it) had been laid. The way was now open for the liquid-fuel experimenters.

2

THE EXPERIMENTERS

Early in 1915, Goddard began his life-long series of experiments by buying firework rockets. He wanted first to measure the efficiency of ordinary rockets, and second to determine how to increase their efficiency. By "efficiency," he meant the ratio of the kinetic (motion) energy of the exhaust gases to the heat energy of the powder. The kinetic energy was calculated from average exhaust velocities obtained by firing the rockets in a ballistic pendulum. Heat energy (calories per gram) was found by burning quantities of rocket powder in a calorimeter. Firework rockets, Goddard discovered, had efficiencies of a paltry 2 percent and exhaust velocities of nearly 1,000 feet per second. Coston lifesaving rockets, developed by U.S. Coast Guard officer Benjamin Coston, did slightly better, with efficiencies of about 2.5 percent and velocities of 1,030 feet per second.

As described in his notebooks, Goddard next built devices for testing "three radical changes" in his experiments. He substituted a high-heat smokeless powder for the gunpowder, carried out the tests in a strong steel firing chamber to allow for higher operating pressures, and used a flared De Laval nozzle to increase the expansion of the exhaust gases. After more than fifty firing experiments, he found he had dramatically increased efficiency to 64.5 percent and exhaust velocities to nearly 8,000 feet per second.

Goddard next sought to fire rockets in a vacuum. Such an experiment would conclusively prove that Newton's Third Law of Motion was valid and that reaction propulsion could indeed work in a vacuum. He also wished to learn if rocket performance was greater when there was no air resistance over the nozzle. For these milestone experiments, loaded firing chambers were inserted in specially made tanks, which were then sealed and pumped empty of air. The rockets were ignited by an electric current leading into the vacuum tanks. As Goddard remarked in his notebooks, "When the gases were fired downwards, the recoil kicked the chambers upward, and the rise was registered by a scratch on a strip of smoked glass."

These experiments, conducted in June and July of 1916, were a great success. First, there was no doubt that rockets would work in the rarified upper atmosphere and in outer space. Second, thrust levels actually increased by an average of 20 percent over those of similar rockets fired in the air. Goddard discovered something else. He calculated that if properly designed, the rocket was a far more efficient heat engine than the internal-combustion (diesel) engine or the "best reciprocating steam engine," though the operating time was short.

Goddard was elated. Here was more evidence that the rocket was the most practical way of reaching space. Based on experimental results, he determined the minimum mass for a theoretical moon rocket, one that used solid fuel and was *unmanned*. He also developed a "multiple-charge" rocket, in which pellets of double-base fuel were automatically fed one by one into the firing chamber by a spring mechanism. This was his practical and immediate solution to the problem of prolonging the burning time of the rocket. He knew that the multiple-charge device was still very far from a moon rocket, but it did offer some realizable rocket improvements and had potential for upper-atmosphere soundings, for weather research, or for military applications.

Goddard made liberal use of Clark University's physics laboratories for experiments, but his meager salary as an assistant professor was inadequate for costly rocket work. In September 1916 he approached the prestigious Smithsonian Institution in Washington, D.C., for a research grant, submitting experimental results that promised a high-altitude exploration rocket capable of going ten times higher than research balloons, which could ascend only twenty miles.

The first of several grants came through, but the outbreak of World War I induced Goddard to turn his attention toward developing an offensive rocket weapon. With the cooperation of the Smithsonian, he entered into an agreement with the U.S. Signal Corps. Goddard was sent to Pasadena, California, to undertake secret experiments. By November 1918, after five months, he had made great progress in developing a series of 5-pound, 7.5-pound, and 50-pound projectiles fired from tube launchers, and had made advances with other, similar projects. His work impressed military observers when he gave demonstrations at the Aberdeen Proving Ground in Maryland. However, a few days later Germany surrendered, and his experiments for the Army came to an end.

The "moon rocket" was not forgotten. Goddard included it in his Smithsonian report *A Method of Reaching Extreme Altitudes,* but only as a theoretical exercise. The public did not take it that way. When the report was first made available to the public in January 1920 (though it had actually been published in December 1919), the press overlooked the dry-as-dust formulas and seized on the moon rocket. Sensationalizing "moon rocket" stories appeared in newspapers across the nation and overseas. For months on end, Goddard was both lampooned and praised; scores of people volunteered for flights to the moon and even to Mars. Goddard, always a retiring person, became even more press-shy and secretive.

Yet unlike Oberth's *Die Rakete,* Goddard's *Method* did not precipitate a space-travel movement. A good deal of publicity was indeed generated, but it was short-lived and largely unscientific; most of the stories greatly distorted and simplified Goddard's theories. Also unlike Oberth, Goddard was not (openly) an advocate of manned spaceflight using liquid-fuel rockets but was still in the solid-fuel phase of his work. The public was at that time totally ignorant of the promise of liquid fuels and of the limitations of solids. As we have seen, Goddard himself was aware of the theoretical superiority of liquids but was very cautious in his assertions. His 1919 report contained only a passing reference to liquid hydrogen and liquid oxygen, buried in a footnote. He stubbornly clung to his solid-fuel cartridge system, despite numerous difficulties in developing the mechanism. "Nozzle split," "chamber cracked," "cartridge jammed," "magazine blew up" were typical entries in his notebooks. Finally, in January 1921, thoroughly exasperated and realizing he was on the wrong track, Goddard did switch to liquids.

How does one start building a liquid-fuel rocket from scratch when this has never been done before and no blueprints exist? Goddard was a practical and well-trained physicist. He knew that first of all he should decide on the specific fuel. He was aware of the high theoretical exhaust velocity of lox-hydrogen (about 17,000 feet per second), and probably knew that the technique for liquefying hydrogen had been developed by James Dewar in the late nineteenth century. But in the 1920s liquid hydrogen was still extremely scarce, being used only in limited quantities for research. Goddard hence chose a mixture of lox and readily available ether, which has a high heat value. Step one was to see how the liquids behaved on contact. The experiments were made with asbestos-wrapped test tubes inserted into or clamped on blocks of wood. The liquids were fed either by lox evaporation or by pressure from carbon dioxide. It turned out that the mixture did not explode on contact but burned smoothly when ignited with electrically heated platinum wire. In more advanced tests, a small combustion chamber with nozzle was suspended on a spring scale to see if the overall principle of a liquid-fuel rocket would work. This minimotor generated four pounds of "lifting force" for a short time without mishap, demonstrating that "sufficient combustion took place."

Goddard's next challenge was to select the right pumps for feeding the liquids into the full-scale combustion chamber. But how was he to make the choice? Nobody had ever pumped lox, at −297 degrees Fahrenheit, quickly into a combustion chamber. There was also the matter of how to handle rapid lox evaporation under high pressure. Both fuel and oxidizer pumps had to be extremely light, rugged, and nonleaking. After low-feed pressure pumps, Goddard tried centrifugal, rotary, piston, and high-speed multivane pumps. Problems arose, as is evident in his notebooks: "it jammed," "it gave no pressure," and so on. The rotary and centrifugal pumps, in particular, leaked. Finally, Goddard met success with simple piston pumps.

Cooling the combustion chamber presented another headache. In December 1922 Goddard thought of adding an extra pump to inject water through a ring of small holes around the top of the chamber. This water, atomized as steam, was to spread evenly around the chamber. But the system failed: the rocket froze up, burned out, and exploded its valves. Then, in March 1923, Goddard became the first to test the important principle of regenerative cooling. One of the propellants (the oxidizer) circulated around the chamber in its own jacket before injection; the fuel then entered through a central hole at the top. However, the regenerative method did not become Goddard's standard chamber-cooling method, and by 1925 he appears to have begun using a heat-resistant, ceramic-coated chamber and nozzle. At the same time, he switched from ether to gasoline as his propellant, and made the significant advance of combining both fuel and oxidizer pumps into a single double-acting unit.

His sponsor, the Smithsonian, grew impatient that no rocket had yet flown, but the inventor was becoming increasingly confident of his design and in October 1925 began construction of a flight vehicle. One important change was that lightweight magnesium alloy and aluminum were substituted for brass and steel. To save additional weight and to simplify the motor, Goddard temporarily abandoned the new and balky pump for a carbon dioxide gas-pressure system. In January he completed the launching frame. Finally, on March 16, 1926, at his Aunt Effie's farm in Auburn, Massachusetts, Goddard successfully tested the world's first liquid-fuel rocket. It was 10 feet long and weighed 11 pounds. Hesitating at first, it built up thrust of about 9 pounds, overcame its weight, then soared 41 feet in 2.5 seconds at an estimated 60 miles per hour. It landed 184 feet away. Only four people witnessed the epochal event: Goddard; Esther, his wife of two years; Henry Sachs, his machinist, who ignited the rocket with a blow-torch on a pole; and Dr. Percy Roope, an assistant professor of physics at Clark, who measured the distances with a theolodite. Goddard described the launch to relatives, but requested the Smithsonian to refrain from making a public statement. Officials at the Smithsonian respected his wish. They believed he was waiting until "a conclusive spectacular demonstration could be made." They waited and waited. As it turned out, the world's first flight of a liquid-fuel rocket remained a virtual secret for exactly ten years—until March 16, 1936, when Goddard's second Smithsonian report, *Liquid-Propellant Rocket Development,* was published. In the meantime, many other researchers tested liquid-fuel rockets and were unaware of the details of Goddard's accomplishments.

Whether secrecy was warranted or not, Goddard's technical achievements until his death in 1945 were brilliant, though problems were endemic. After his second liquid-fuel flight on April 3, 1926, he decided to build a rocket that was twenty times larger in propellant capacity. This vehicle, incorporating many new features, including multiple injectors and flow regulators, was completed in December. A new launch tower was built. No flights were made, however, because of technical difficulties. In a static test in May 1927, the rocket attained a thrust of more than 200 pounds, but the gasoline tank exploded. After a similar mishap in August 1927, Goddard decided that this rocket was too ambitious and began designing an entirely new vehicle one-fifth the

Figure 4. Robert Goddard standing beside the world's first successfully flown liquid-fuel rocket, which was launched on March 16, 1926. Note the rocket nozzle on the top. The asbestos-covered cone on the bottom directed exhaust gases away from the propellant tanks below. Goddard found this "nose-driven" design unstable and later shifted the rocket motor to the bottom. The "tail-driven" configuration was more stable and became standard in all rockets.

size of the 1926 model. Variations of this rocket, with simpler, replaceable parts, improved injectors, regenerative cooling, vanes, and gyroscopic stabilization, made Goddard's third and fourth flights in 1928 and 1929.

The latter flight tested the first instrumented liquid-fuel rocket, which carried a barometer, a thermometer, and a camera for taking photos of the dials. It went up 90 feet, then crashed in a terrific explosion, without deploying its parachute. People in the vicinity thought that an airplane had crashed; and once again, Goddard's work generated sensational headlines. In 1930, to escape the publicity, he shifted his entire rocket operation to the remote town of Roswell, New Mexico. The new location had the additional advantages of being able to provide him with wide-open spaces and excellent climate year-round for testing.

With the enthusiastic support of the celebrated aviator Charles Lindbergh, Goddard obtained additional research funds from the Guggenheim Foundation for the Promotion of Aeronautics. His years in Roswell, from 1930 to 1932 and from 1934 to 1941, were extremely productive. His rockets became increasingly sophisticated. Among the numerous milestones achieved were streamlined casing and remote control (1931), gyroscopically controlled air vanes (1932), a rocket containing a cluster of four combustion chambers (1936), a pendulum stabilizer (1935), naphtha fuel (1936), a lightweight, wire-wound pressure storage tank (1937), catapult launching (1937), gimbal steering (1937), high-speed centrifugal rocket pumps and a gas generator for the turbines (1938), and double turbines (1938). Goddard flight-tested thirty-one rockets in New Mexico, the most successful reaching 7,500 feet on May 31, 1935; another, on March 8, 1935, attained over 700 miles per hour. A static test on July 17, 1941, produced 825 pounds of thrust (3,040 horsepower) for 34 seconds.

In 1941, with war impending, the U.S. military became more interested in rocket power. At about this time, a friend told Goddard that he could progress faster by being less "secretive." Both factors persuaded Goddard to offer his services to help the Navy develop one of its JATO (Jet-Assisted Takeoff) projects, aimed at shortening the takeoff distances of heavily loaded seaplanes through the assistance of liquid-fuel rockets. He worked on this from 1941 until his death in 1945. The project was not commensurate with his enormous talents, though JATOs were then critically needed and some of the work led to the throttable Curtiss-Wright XLR25-CW-1 rocket engine ("throttable" means that the thrust can be raised or lowered, in this case with a pilot-operated control), which powered the supersonic Bell X-2 after the war. (Goddard was to have joined Curtiss-Wright after the armistice but died August 10, 1945, of throat cancer.)

During his lifetime, Goddard was issued 48 patents covering basic rocket hardware. After his death, his widow obtained an additional 131 posthumous patents, for a total of 214. In 1960 the National Aeronautics and Space Administration (NASA) acquired the use of these patents at a cost of one million dollars. The question of what role Goddard played in "mainstream" U.S. rocketry and whether he merits the title "Father of U.S. Rocketry" is thus moot. Yet although he worked far from the public eye for much of his lifetime, there is no disputing that his determination and vision served to inspire countless researchers in the decades following.

■ THE ROCKET SOCIETIES

Every international movement produces its clubs of zealots and core workers. The same was true of spaceflight when it became the rage in the early 1920s.

The earliest space-travel group was the Soviet Union's Society for the Study of Interplanetary Travel, a division of the Moscow Society of Amateur Astronomers. It was founded in 1924 by Fridrikh A. Tsander, a Latvian, who had become fascinated with spaceflight when he had read Tsiolkovsky's works as a child. (Tsander himself became a leading figure in Soviet spaceflight circles from World War I to the early 1930s, but was virtually unknown in the West at the time.) The society's history is difficult to trace, though it is known that the group briefly became part of the Military Science Division of the N. E. Zhukovsky Air Force Academy and consisted of about 200 members. Among its aims was to unite all people in the USSR working on the spaceflight problem, to obtain information on the subject from the West, and to engage in research, particularly regarding the military applications of the rocket. Tsiolkovsky was asked to participate, but pleaded old age and was made an honorary member. In addition to experiments, the society planned to start a journal. Neither project materialized. The group, also called the Society for the Study of Interplanetary Communication (abbreviated OIMS in Russian), became a short-lived debating club. Its most spectacular performance was a heated, three-day debate over a shameful and erroneous report claiming that Goddard had sent a rocket to the moon on August 5, 1924. Posters were printed, and the horse militia was called out to keep order in the streets around the auditorium. Perhaps this episode dealt a mortal blow to the society's credibility, for it collapsed less than a year later.

In the West, the earliest space group was the Österreichische Wissenschaftliche Gesellschaft für Hohenforschung (Austrian Scientific Society for High-Altitude Exploration), founded in Vienna in 1926 by Dr. Franz von Hoefft and Baron Guido von Pirquet, who became outstanding names in early space-travel theory. As a result of personality clashes and methodological disputes, Oberth was excluded from the membership. As was the case with OIMS, experimentation was made impossible by lack of funds, though the Austrians set an example for the so-called German Rocket Society, which became the largest and most important of the space-travel groups.

The Verein für Raumschiffahrt (Society for Spaceship Travel, or VfR), popularly called the German Rocket Society, was founded July 5, 1927, in an alehouse in Breslau (now Wrocław, Poland). Max Valier suggested that the group raise money to finance Oberth's rocket work. Valier himself declined the VfR presidency because of heavy speaking commitments, and the position was assumed by Johannes Winkler, an engineer and church administrator. Winkler also assumed editorship of the VfR's journal, *Die Rakete (The Rocket)*, the first magazine devoted to promoting spaceflight. It became an invaluable forum for presenting ideas on this topic. For example, von Hoefft wrote a series proposing the development of a number of rockets, ranging from a simple balloon-borne lox-alcohol sounding rocket called the Hoefft Rocket 1 (or RH1), which would ascend 100 kilometers (62 miles), to an advanced staged version (the RH7) designed to reach the moon, Mars, or Venus. Von Pirquet published an essay in which he worked out trajectories to the planets. Another interesting contribution, published in 1929 by Franz Abdon Ulinski, was an idea for an electron-powered spaceship. The idea was impractical, since it defied laws of physics (Ulinski planned to capture and recycle the ejected electrons that supplied the motive force), but the electron rocket was still a precursor to solar-electric propulsion. *Die Rakete* boosted VfR membership to

Figure 5. "Electron rocket" of Franz Abdon Ulinski of Austria, 1927.

about one thousand, but the journal ceased publication in 1929 because the society decided to devote its limited resources to experimentation.

Winkler had already begun independent experimentation in 1928, and had published the solid-propellant thrust curves he had obtained through his work at the machine shop at the Breslau Technische Hochschule (Breslau Technical High School). He soon progressed to liquids and received financial and technical support from VfR benefactor Hugo A. Hückel, a wealthy manufacturer. On February 21, 1931, the lox-methane Hückel-Winkler 1 (or HW-1) was successfully launched near Dessau, at Gross Kuhnau. It reached an altitude of about 2,000 feet, and was hailed as Europe's first successful liquid-fuel rocket. In fact, it was thought to be the first liquid-fuel rocket to be tested anywhere, since Goddard's previous five flights had all been kept secret. (In addition, German pyrotechnician Friedrich Sander claimed to have secretly tested a liquid-fuel rocket on April 10, 1929, using gasoline and an undisclosed oxidizer.) In any case, the fundamentals of the liquid-fuel engine had been clearly established by 1930.

Oberth, who had been hired as a technical adviser for the influential silent science fiction film *Frau im Mond (Woman on the Moon),* constructed a demonstration lox-gasoline motor called a Kegeldüse (cone-jet), which was static-fired on July 23, 1930. It produced 15.4 pounds of thrust for 90 seconds. (Oberth was to have launched the rocket on the day of the film's premiere, but because of technical difficulties this never came about and only a static test was conducted.) One of the witnesses of this test was Franz Ritter of the Chemische-Technische Reichsanstalt (the Government Institute for Chemistry and Technology), who certified the motor. Since the Reichsanstalt was an important agency, much like the U.S. Bureau of Standards, the certification was influential in convincing critics of the viability of the liquid-fuel rocket engine and in drumming up support for the VfR.

Also in 1930, the VfR moved to Berlin and founded its Raketenflugplatz (Rocket Airfield). Earlier in the year, at Bernstadt, Saxony, several members had already begun experimenting with a small lox-gasoline Mirak (Minimum Rocket), in which carbon dioxide cartridges, normally used to pressurize seltzer-water bottles, served to inject the fuel. This simple device yielded surprisingly powerful thrusts: initially 14 ounces, later up to 7 pounds. Cooling was arranged by placing the combustion chamber in the lox tank. However, there were no safety valves, and explosions were frequent. The establishment of the Raketenflugplatz promoted rapid progress, though economically these were difficult times and the researchers devoted much time to ''scrounging'' for free materials. One space-struck idealist was teenager Wernher von Braun, who conducted rocketry experiments whenever school was not in session, and especially on weekends.

Mirak gave way to Mirak II and Mirak III, the latter capable of yielding 70 pounds of thrust for 30 seconds. The final series consisted of the Repulsors, numbered I to IV. None of the VfR rockets seem to have used pumps; the propellants were injected by means of either inert nitrogen or lox-evaporation gas pressure. Cooling was usually achieved by using a water jacket. Interestingly, the last Repulsors switched to lox and watered alcohol as propellants. But before the VfR finally dissolved in the winter of 1933 due to internal conflicts, several semi-official rockets (developed by the society's so-called Project Magdeburg) were tried, some with regenerative cooling. Yet VfR experimental data are very scarce. Altogether, hundreds of static tests and roughly a hundred flights were made, though top altitudes barely exceeded 3,000 feet. Average thrust was about 100 pounds, with a maximum of 550 pounds for the Magdeburg rockets.

What was gained from the VfR experiments? Documents show that they produced few technical innovations, but they without doubt gave invaluable experience to future researchers in rocketry, the most prominent of whom was von Braun. The German Army took an active interest both in the highly publicized experiments of the VfR and in those of its more competent members, such as von Braun. Even before Hitler came to power in 1933, some people in the German Army had begun working toward rearmament, despite the armament restrictions imposed upon Germany by the French in the Versailles Treaty following World War I. Liquid-fuel rockets appeared to be a highly promising weapon for the new, secret Army; and rockets were not explicitly covered in the treaty. Thus, in the spring of 1932 the Raketenflugplatz received visitors in business suits, eagerly asking for details of rocket construction and performance. These men were members of the small team of Army officers charged with rocket development. There were further visits, as well as talks with von Braun and others. Finally, in the fall of 1932, von Braun became the first VfR member hired to work in the Army's new secret rocket program. Others soon followed. Thus began the research leading to the dreaded V-2 rocket of World War II.

With regard to his motives in offering his services to the Army, von

Braun later observed that he and other VfR members simply had no idea in 1932 where their rocketry activities would lead, and that their Army employment predated Hitler's rise to power. Von Braun was also "sure" that the Raketenflugplatz "was utterly inadequate to even begin the vast experimental program which must be the precursor of success. It seemed that the funds and facilities of the Army were the only practical approach to space travel." Some people regard von Braun as an apolitical visionary with outstanding gifts in engineering and organization who seized the only opportunity available to pursue his single-minded goal of developing a space rocket (with the added advantage that he could obtain his doctorate in physics through his Army work on rockets); furthermore, during the war years he was doing his duty for his country. Others suspect he knowingly chose to ignore the moral implications of his work and fully realized the consequences of making rockets for the Army. We may never know the truth.

Second to the VfR in importance was the American Interplanetary Society (AIS), started in New York City in 1930 by a group of science fiction authors. At first, its members did little but hold lectures and mimeograph bulletins. Then, early in 1931, AIS vice-president G. Edward Pendray visited Europe and toured the Raketenflugplatz. Pendray resolved to begin experimentation at home, based on what he had seen in Germany. By the summer of 1932, the first AIS rocket was under construction. Costing all of $49.40, it was no feat of engineering but a model of thrift and ingenuity: it had a tin-sheet body with wooden fins, an aluminum water jacket that had formerly been a cocktail shaker, a parachute holder made from an aluminum saucepan, and a silk pongee parachute bought in a department store. On November 12, 1932, the lox-gasoline rocket was static-fired in a field near Stockton, New Jersey. It registered a thrust of 60 pounds for 20 or 30 seconds on a spring scale, but after a further test the motor was not considered fit to fly. Not until May 14, 1933, did an AIS rocket lift off. It went up 250 feet. The second and last AIS rocket flew September 9, 1934, to a respectable 1,338 feet. After this, the group (now renamed the American Rocket Society or ARS), devoted its limited resources toward more practical static tests in which rocket engines were fired on stands and closely observed, so that data could be collected. (From the start, Goddard was asked to assist the AIS researchers but remained aloof. He considered them amateurs, which indeed they were, but later regarded them with more respect, after more engineers joined the society.)

The ARS's static-testing period, from 1935 to 1941, proved fruitful. The experimenters' chief concern was to find an optimum means of cooling the motors for long runs. On ARS Test Stands 1 and 2, a variety of techniques were tried: standard water jackets, "heat sponge" aluminum blocks, heat-resistant nozzles of nichrome or other metals. Then came a breakthrough. James H. Wyld, who read German, learned of the regenerative-cooling principle in a paper by the Austrian scientist Eugen Sänger. (Wyld was unaware that Tsiolkovsky and Oberth had discovered the principle earlier, or that Goddard and the VfR had actually tested it.) He constructed a rocket incorporating a regenerative cooling system, and tested it on ARS Test Stand 2 in 1938. After a hot firing, Wyld's little motor was, he found, almost cool to the touch; the system was clearly a success. In December 1941, shortly after the bombing of Pearl Harbor, Wyld and three other ARS members (John Shesta, Lovell Lawrence, Jr., and H. Franklin Pierce) formed America's first commercial liquid-fuel rocket company, Reaction Motors, Inc. Using the regenerative-cooling principle, the company made major contributions to rocketry. The most famous was a four-cylinder engine generating 6,000 pounds of thrust. Known as the 6000C4, it was used in 1947 to power the Bell X-1, the first plane to break the sound barrier.

The ARS continued to flourish, becoming the largest professional rocket engineering society in the United States. In 1963, when it merged with the Institute of Aerospace Sciences to become today's American Institute of Aeronautics and Astronautics (AIAA), it listed 20,000 members, the elite of the American aerospace industry.

In the Soviet Union the OIMS had folded in 1924, but new rocketry groups sprang up and enjoyed remarkable success. Among them was a laboratory begun in Moscow in 1921 to develop Nikolai I. Tikhomirov's smokeless-powder war rocket. It moved to Leningrad in 1927 and was renamed the Gas Dynamics Lab. Researchers affiliated with the lab at first continued solid-fuel studies; but on May 15, 1929, at the suggestion of member Valentin P. Glushko, they formed a subdivision (known as Department II) to study liquid-fuel and electric rocket engines. Department II, headed by Glushko, was to play a critically important role in the history of spaceflight.

Although the Gas Dynamics Lab operated under the control of the

Soviet Army, Glushko managed to promote the development of an electric rocket engine producing very small thrust yet high specific impulse. This motor had no foreseeable military applications but theoretically was ideal for use in deep space. Perhaps, like von Braun, Glushko took orders from the military but was secretly an ardent spaceflight enthusiast, working any way he could toward achieving spaceflight, which he considered his true priority. (In 1926, at the age of eighteen, Glushko published an article entitled "Station beyond Earth," one of the first essays ever written on artificial satellites and space stations.) However, the Gas Dynamics Lab concentrated primarily on liquid-fuel JATOs and missile powerplants. Between 1930 and 1933 it developed an impressive array of engines, ranging from the OPM and OPM-1 (Experimental Rocket Engine) to the OPM-52, which was capable of generating 660 pounds of thrust. Achievements included new oxidizers (mainly nitric acid-based) and fuels, including hypergolics; bell-shaped, high-thrust nozzles; gimbaled engines; and turbopumps. The later history of the laboratory (and of Glushko) is sparsely documented, though it is known that in 1933 it merged with the Group for the Study of Reactive Motion (GIRD) and became the Rocket Research Institute (RNII). The RNII produced engines up to OPM-102 in 1937 and 1938—years of the terrible Soviet purges in which many intellectuals were exiled or murdered on false charges of treason. Victims included an unknown number of rocket scientists.

There had actually been several GIRDs, the most important of which was the Moscow Group for the Study of Reactive Motion (MosGIRD), formed in 1931 and led by the brilliant researcher Sergei P. Korolev. MosGIRD started as a local civilian experimental club but eventually became the "central" GIRD. On August 17, 1933, it launched the USSR's first hybrid (liquid-solid) rocket, called the GIRD-09, and on November 25, 1933, a true liquid-fuel rocket, the GIRD-X. In the following year the organization was absorbed into the RNII and placed under the jurisdiction of the military.

Korolev, sometimes wearing Army officer's pips, was subsequently assigned the task of designing missile engines, boosters, and experimental rockets. Like Glushko, however, he had always been primarily interested in spaceflight and rocket planes (though not Shuttle-type craft). The 1937–1938 purges did not leave him unscathed. He was accused of treason, arrested, and sent to forced labor in Siberia's Kolyma gold mines. The Soviet rocketry and spaceflight movement was enduring its blackest days. It is thus not surprising that Soviet histories of rocketry record a sudden falling off of rocketry activities, though a few JATO and rocket airplane projects lingered until the eve of the war. One Soviet source says that in 1939 Korolev was unable to witness the first flight of one of these prototypes, "for reasons beyond his control."

■ EUGEN SÄNGER

The Space Shuttle concept also originated in the 1920s and 1930s. Oberth exchanged acrimonious words with Max Valier over Valier's insistence that reusable spacecraft were superior to "expendable" (one-time only) rockets. Valier's idea had merits but was incomplete. He believed that spaceflight would be achieved through the gradual evolution of the Junkers G-23 propeller-driven airplane. The first step would entail the addition of two auxiliary rocket engines for testing the "jet propulsion" principle. If this worked, step two would involve installing four rocket engines, with an auxiliary propeller engine as a back-up. The stratospheric version had six rocket engines and a pressurized cabin. "The final stage," said Valier, "is to develop a rocket ship which will ascend vertically from a launching tower, and·which will be . . . the spaceship at last."

Eugen Sänger of Austria, who held a doctorate in aeronautical engineering from Vienna's Technische Hochschule, focused on the rocket-propelled stratospheric plane. He believed that such a craft was within the range of existing technology and that the next step was an Earth-orbital space plane for the construction, transport, and supply of a space station; development of interplanetary and interstellar spaceships would follow. The basic rocket plane, which he dubbed the "Silver Bird," was to occupy him for more than thirty years.

Sänger had become interested in spaceflight when, at the age of thirteen, he had received as a present Kurd Lasswitz's 1897 novel *Auf zwei Planeten (From Two Planets),* in which Martians come to Earth in a gravity-nullifying ship. Sänger first seriously thought of space travel in 1924, upon reading Oberth's *Die Rakete.* The book inspired him to take up the study of aeronautical engineering, and he earned a university degree in the subject in 1929.

From the start, Sänger favored the reusable rocket plane over "ballistic'' systems, though the former entailed many more technical difficulties: working out the aerodynamics and trajectories, designing a safe, reusable engine, overcoming reentry heating, effecting recovery, and so on. He knew that the engine was the key to a workable system and therefore concentrated upon propulsion; cooling the reusable engine for long flights was of paramount importance. In 1933 he published his theories in a book entitled *Raketenflugtechnik (Rocket Flight Technology),* another seminal work of astronautics. Sänger also designed a comprehensive research program. Among his objectives was to find heat-resistant materials for lining the combustion chamber and nozzle throat, and a suitable propellant coolant, though he had already proposed regenerative cooling.

In 1934, at the Technische Versuchsanstalt (Technical Research Institute), he began conducting experiments on a variety of subjects, including regenerative cooling. He obtained exceptional burning times of 15–20 minutes, and one of half an hour. Cooling jackets consisted of coils wrapped around the combustion chamber. Propellants were light diesel oil and gaseous oxygen. Thrust levels were small and measured on a spring dynamometer. In October 1934, after 135 tests, Sänger had to cease his experiments because the noise bothered nearby residents. Nevertheless he had collected a wealth of data, and on February 9, 1935, he was granted Austrian Patent 144,809 for regenerative cooling and other ideas. He was also granted patents in other countries.

From 1937 to 1942 at Trauen, Germany, Sänger continued research on a much larger scale for the Deutsche Versuchsanstalt für Luftfahrt (German Research Institute for Aviation). He conducted many static tests with a high-pressure regeneratively cooled engine yielding 2,200 pounds of thrust, and drew up plans for one that would yield 220,560 pounds of thrust; the chamber for the latter was actually built. Experiments progressed to the point where wind-tunnel models were tested. To spare the project from possible wartime cancellation due to shifts in priorities, the "Silver Bird" was renamed the "Antipodal Bomber" and assumed a new role. It became an Earth-orbiting supersonic bomber with a gross weight of 100 tons. The project required a great deal of mathematics, on the part of both Sänger and his assistant Irene Bredt (later, his wife). After its tremendously accelerated boost by rocket into the upper atmosphere (to, say, an altitude of 155 miles in about 5 minutes), the Antipodal Bomber coasted down at a certain angle like a roller coaster and ricocheted or bounced off a layer of dense atmosphere (say, at an altitude of 25 miles), then shot up again (unpowered), bounced off another upper layer (at about 80 miles), coasted down again, and continued this pattern for great distances around the globe. The Bomber (sometimes known as the Sänger-Bredt Antipodal Bomber) thus flew in a gradually diminishing wavelike trajectory—again, like a roller coaster.

But wartime exigencies, including fuel shortages, did catch up with the project and forced its cancellation. Ater the war, as leader of the newly formed International Astronautical Federation, Sänger became a tireless advocate of the spaceplane. He died in 1964, before he could see his Silver Bird take flight, greatly modified in the form of the Space Shuttle.

3

THE V-2 ROCKET

T he V-2, the world's first large liquid-fuel rocket and one of World War II's most awesome weapons, began as a textbook exercise in ballistics. In 1926 a young German artillery officer named Karl Emil Becker, who was studying under Professor Julius Cranz, helped his teacher write what was to become a well-known standard: *Lehrbuch der Ballistik (Textbook of Ballistics)*. The work contained a lengthy section on rocketry (probably contributed by Becker), a discussion of Goddard's 1919 paper, and an analysis of the spaceships proposed in Oberth's *Die Rakete*. In the course of this project, Becker conceived a lifelong interest in rocketry. Three years later, now a colonel (he would eventually be promoted to general) and chief of the Army's Department of Ballistics and Munitions, Becker ordered a thorough examination of the literature, the first step toward adapting liquid-fuel rockets for military needs.

The engineer assigned this task, Captain D'Aubigny von Engelbrunner Hörstig (also known simply as von Hörstig), quickly reached a dead end. Aside from Oberth's stunt rockets for the movie *Frau im Mond,* there had been almost no work done in Germany on liquid-fuel rockets. Universities and private industry had thus far shown no interest in the technology. The sole exception was the Gesellschaft für Industriegasverwertung (Association for the Utilization of Industrial Gases), led by Paul Heylandt, which in 1929 began experimenting with gaseous carbon dioxide as a fuel for Max Valier's rocket cars.

Then, in 1930, von Hörstig was joined by Captain Walter R. Dornberger, who became head of the program. Technical progress quickened rapidly thereafter. On April 19 Valier first tested a liquid-fuel version of his rocket car, called the Rak 7, using a nitrogen-fed mixture of lox and watered alcohol, but he was killed a month later when the same vehicle, operating on a mixture of lox and watered paraffin, exploded. In June the VfR initiated experiments with its Mirak I; in July Oberth successfully tested his Kegeldüse; and in September the Raketenflugplatz was founded. Despite Valier's death, it was obvious that liquid-fuel rockets worked. This realization resulted in a pivotal meeting on December 17, 1930, attended by Becker and Dornberger and led by Ordnance Colonel Erich Karlewski, at which the equivalent of $50,000 was granted for the Army's own rocket program. Comparable sums were allotted in subsequent years.

"You have to develop a liquid rocket which can carry more payload than any shell presently in our artillery [and] . . . farther than the maximum range of a gun," read Becker's order to Dornberger. "Secrecy of the development is paramount." The first step was to recruit experts (preferably experimenters) in liquid-propellant rockets. In 1930 this was "no small task," Dornberger later recalled. At that time rocketry was, in his words, "a sphere of activity beset with humbugs, charlatans, and scientific cranks, and sparsely populated with men of real ability." The VfR showed promise, but its members were for the most part amateurs. More professional was Heylandt's Association for the Utilization of Industrial Gases. In 1931 Dornberger contracted this firm to develop a small liquid-fuel motor for "basic experiments." It was a steel, cylindrical, double-walled (probably regeneratively cooled) engine of 45 pounds thrust. The engine, with the test rig, was turned over to the research branch of the Army's Department of Ballistics and Munitions, where it was tested with different propellants under the direction of chemist Kurt Wahmke. Working with a pyrotechnist and with his own students, Wahmke obtained useful data, until a fatal explosion in March 1934 ended this phase of the experiments.

Meanwhile, Dornberger was slowly assembling a competent technical staff. In his visits to the Raketenflugplatz he had been unimpressed with the rockets but had been "struck" by the exceptional abilities of Wernher von Braun. On October 1, 1932 (or November 1, according to other sources), von Braun was signed as Dornberger's first technical assistant. The second was another VfR member, Heinrich Grünow. And the third was Walter H. J. Riedel (no relation to the VfR's Klaus), who had assisted Valier at the Heylandt plant. A rocket research facility had also been set up at an old army firing range near Kummersdorf, about seventeen miles south of Berlin. This was somewhat inappropriately called Experimental Station West. Here, Dornberger was responsible for making both solid- and liquid-fuel barrage rockets. A second liquid-fuel motor, running on a mixture of lox and alcohol and capable of producing 650 pounds of thrust for 60 seconds, was obtained from the Heylandt company and was tested at Kummersdorf on December 21, 1932.

This test nearly ended in disaster. The stand contained an elaborate assortment of dials and gauges to measure engine-flow rates, pressures, fuel-mixture ratios, and so forth, but the ignition system was crude: von Braun held a long stick with a can of gasoline fastened to the end, and shoved this under the exhaust nozzle as the propellants were let out. A tremendous explosion resulted, destroying the engine and stand but miraculously hurting no one. The men learned quickly from their mistakes, and by mid-1933 a more advanced version of the motor was made. (The following year Dornberger hired one of the motor's designers, Arthur Rudolph, who had been with Valier in his final moments.) This second-generation Heylandt engine was installed in the initial flight vehicle, called the Aggregat-1 (or A-1), which was 4.5 feet long and a foot in diameter. However, after various checks it was found that the A-1 was too nose-heavy and the experimenters decided to proceed at once to the next stage. This was the A-2.

The dimensions of the A-1 and the A-2 were similar, but in the latter the gyroscopic stabilizer was closer to the rocket's center of gravity. Two models were built, affectionately called Max and Moritz after the mischievious "Katzenjammer Kids" comic strip. In late December 1934, at Borkum Island in the North Sea, both "kids" flew beautifully to an altitude of about 6,500 feet. This success led to increased funds, more staff, bigger motors (up to 3,300 pounds of thrust), and plans for moving to a more spacious, self-sufficient facility. Meanwhile, the German Air Force became interested in rocketry, specifically for use in JATOs, rocket aircraft, and missiles. Thus was born the idea of a joint Army–Air Force rocket establishment under Army control, later variously called the Heeresversuchsstelle Peenemünde (Army Experimental Station at Peenemünde), the Heeres-Anstalt Peenemünde (Army Establishment at Peenemünde, abbreviated HAP), or simply Peenemünde. Officially established in April 1937, Peenemünde was ideally situated on the wooded island of Usedom, in the Baltic. It thus had two important advantages: a remote location and a long test range extending out over the sea.

It took two years and millions of Deutschemarks to build Peenemünde. By late 1935 Kummersdorf's staff had grown to 80, which mushroomed to 300 when the move was made to Peenemünde. Included were other former VfR members. In the same period, the A-3 rocket was developed, producing 3,300 pounds of thrust for 45 seconds, and having a height of 21.3 feet and a diameter of 2.2 feet. It featured three-dimensional gyroscopic control, molybdenum exhaust vanes, a radio link for fuel-flow cutoff, magnetic servo valves (valves activated by small electric motors), and liquid nitrogen pressurization. However, when test-fired in December 1937, all four A-3 vehicles failed after good lift-offs. According to wind-tunnel calculations, the gyroscopic system was too weak and the exhaust-vane movements too slow to counteract wind. Yet by March 1936, even as the A-3 was being constructed, the basic parameters of the A-4 had been set. And it was the latter vehicle that would eventually achieve fame under the designation V-2.

Veteran artilleryman Dornberger worked out the ultimate size of the rocket with von Braun and Walter ("Papa") Riedel, later Peenemünde's chief designer. Dornberger's idea was to double the 80-mile range of the Paris Gun used in World War I. The planned payload was one ton of high explosives. The rocket was designed to be transportable by rail and had to be able to pass through railroad tunnels; interestingly, it was this fact which determined the A-4's fin size. The length was projected at 45 feet, the diameter at 5 feet, the weight at 12 tons, and the thrust at 50,000 pounds for 60 seconds. When the A-3 was abandoned, engineers went ahead with plans for the A-5, which would use the A-3's well-tried motor. The distinguishing feature of the A-5 was to be a more powerful guidance system. Like Goddard's unsuccessful 1929 rocket, the A-5 carried a barograph, a thermograph, and a movie camera to record the readings. When Adolf Hitler paid a visit to Kummersdorf on March 23, 1939 (his only first-hand look at rocket development), he was shown the A-5 by von Braun and also witnessed some static tests; but he failed to grasp the potential and significance of the work. In fact, the A-5 proved to be an invaluable recoverable (that is, with parachutes) research tool which helped perfect the guidance systems, servo motors (small electric motors for controlling a variety of subsystems), and graphite vanes incorporated in the A-4. The A-5 achieved stable flights with ranges (that is, horizontal distances) up to 11 miles, and had a peak altitude of 7.5 miles.

When Hitler visited Kummersdorf in 1939, the A-4 vehicle was already under construction. Undoubtedly, the biggest problem was how to produce a more powerful engine. Von Braun and Riedel made many contributions to this project, but the crucial engine design was mainly the work of chemist Walter Thiel, who had joined Kummersdorf in the fall of 1936. Thiel began with a motor yielding 3,000 pounds of thrust, then built a model yielding 9,000 pounds, and by 1939 had completed the basic model of a compact, high-pressure (750 pounds per square inch) engine with a thrust of 50,000 pounds. The double-walled (regen) motor burned a mixture of lox and watered alcohol, which helped cool it further. But internal "hot spots" (heat-sensitive areas) still existed. Thiel's colleague Moritz Pöhlmann suggested an innovative solution called "film cooling," in which the inner wall of the combustion chamber was sprayed with a fine film of alcohol. Hot spots were drilled with holes, then filled with Wood's metal (a low-melting-point alloy), which melted under the flames and allowed the cooling alcohol to enter. Dornberger suggested the further refinement of using centrifugal injectors to atomize the fuel, resulting in efficient combustion and a high exhaust speed. Von Braun proposed arranging eighteen of Thiel's combination injection-mixing heads (used on the 3,000-pound-thrust motor) in two concentric circles, and this worked out very well.

Von Braun also tackled an important feature of the full-scale 50,000-pound-thrust motor, namely the pumps. These had to be light, yet capable of almost instantly forcing liquid gas (lox) and fuel, each at 300 pounds per square inch, at a rate of 50 gallons or more per second, into the combustion chamber. Von Braun approached pump manufacturers and was surprised when told that these parameters were similar to those for firefighter pumps. Firefighter centrifugal pumps thus became the basis for those in the A-4. Thiel was well aware of the huge amounts of steam generated by hydrogen peroxide when exposed to a catalyst (since about 1935 he had been considering peroxide as a potential A-4 fuel) and logically chose this substance to provide instant ignition for the big rocket's massive pumps, using potassium permanganate as the catalyst.

The final version of the A-4 was a bullet-shaped vehicle 46.9 feet long and 5.4 feet in diameter. It had a take-off weight of 28,229 pounds, a total warhead weight of 2,310 pounds (amatol high explosive), and a thrust of 59,500 pounds for 68 seconds, which propelled it at 3,500 miles per hour maximum to a 190-mile range and a peak altitude of 60 miles. The first two launches, on June 13 and August 16, 1942, were failures. But on October 3, the A-4 succeeded flawlessly. It attained a 125-mile range and came within 2.5 miles of its target, reaching a top speed of 3,300 miles per hour and a peak altitude of 60 miles, or the fringes of space. That evening, Dornberger toasted the event as a decisive one in the history of technology. It marked, he said, "a new era in transportation: that of space travel."

But this was wartime. "Our most urgent task," Dornberger reminded his staff, "can only be the rapid perfecting of the rocket as a weapon." Some 23 months and approximately 65,000 technical alterations after its maiden flight, A-4 became operational. The project was transferred from Peenemünde, and also from university laboratories and private industry (where much contract research and development work was undertaken), to the labyrinthine underground factories of Mittelwerk in the Harz Mountains. Here, the A-4 was mass produced by thousands of slave laborers and other workers. (Only about 250 developmental A-4s were built at Peenemünde.) On July 7, 1943, Dornberger and von Braun again met with Hitler, this time at the latter's headquarters in East Prussia. The Führer was shown models and a film of the first successful flight; his former indifference to rockets vanished. The A-4 was given the highest priority and was dubbed the V-2 (Vengeance Weapon 2), complementing the shorter-range, nonrocket (pulsejet-powered) V-1, which the British came to call the "Doodlebug."

Meanwhile, at Peenemünde, Ludwig Roth of the Projects Office, von Braun, and others were designing and building more advanced models. One that was nearly completed was the A-9: a gliding, winged (or rather finned) V-2 with a maximum horizontal range of about 275 miles at Mach 4.4 (that is, 4.4 times the speed of sound, or 3,500 miles per hour). Peenemünde's wind-tunnel studies on the A-9 were begun in 1940 but discontinued in 1943. Finally, on January 24, 1945, another finned version, originally called the A-10 but redesignated the A4b, successfully climbed to an altitude of 48 miles at 2,700 miles per hour, then glided down before crashing. Drawings for a manned, Shuttle-type A-9 existed, but no further A4b tests were made. In 1940 Arthur Rudolph's wind-tunnel team at Peenemünde, the best-equipped in Germany, also calculated the double-staged A-9/A-10, a massive vehicle

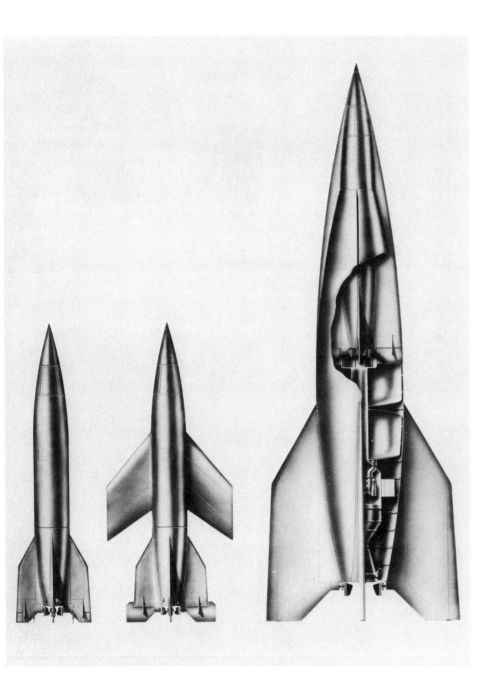

Figure 6. The V-2 family. Left to right: the V-2 (A-4), the A4b, and the A-9/A-10 concept.

eighty feet long. Walter Thiel designed its first stage, which comprised six A-4 engines using a common nozzle and which produced a thrust of 440,000 pounds (later it was redesigned as a single chamber). The upper stage was an A-9. The visionaries at Peenemünde actually contemplated sending the A-9/A-10 to attack a target 3,000 miles away, namely New York City. And they ventured even further. They designed a three-stage A-11 vehicle (its first stage alone was to yield 3.5 million pounds of thrust) that was to be capable of orbiting the Earth, and an A-12 for sending up a satellite. The latter was to be the first step in the eventual construction of a space station.

Back on terra firma, Germany's military situation was grim. By January, 1945, the Allies were converging on the country from both east and west. Some 6,400 V-2 rockets had been fired operationally, but they were introduced too late to have a decisive influence on the war. In April, Dornberger was ordered to evacuate Peenemünde, as U.S. and Soviet tanks advanced upon it. The facility's archives, rockets, and equipment were hidden, but were later captured by American forces. On the second of May, Dornberger, von Braun, and other top staff members at Peenemünde surrendered to the Americans.

■ AMERICAN V-2 ROCKETS

The United States acquired the elite of German rocket scientists, consisting of 118 specialists. (This was a mere fraction of the personnel involved in rocketry development in Germany: at war's end, 4,325 people were employed at Peenemünde and 10,000 were working at Mittelwerk.) Under the direction of Colonel Holger N. Toftoy, chief of U.S. Army Ordnance Technical Intelligence in Europe, 100 operable rockets, numerous rocket components, and tons of scientific documents were shipped from Germany to the United States. In March 1945 one of these rockets was examined by Robert Goddard, who found the experience traumatic. "I don't think he ever got over the V-2," observed a colleague. "He felt the Germans had copied his work and that he could have produced a bigger, better, and less expensive rocket, if only the United States had accepted the long-range rocket." In truth, von Braun, Dornberger, and other German researchers claimed that because of wartime restrictions, they had never seen any of Goddard's patents, much less used them. More important, when one considers

Peenemünde's vast resources and the many scientists who worked there, it is hardly surprising that Goddard and the Germans should have devised similar solutions to problems of rocket engineering. As for the long-range rocket, U.S. authorities may indeed have been apathetic toward or unaware of such a development; but in fact Goddard himself never promoted the idea and was secretive about his work. Thus, although the United States made thousands of wartime rockets, most of these were small, double-base, solid-fuel types like the Bazooka.

Yet shortly before the war ended, the U.S. Army did realize that it was lagging behind in missile development, and on November 15, 1944, Project Hermes was initiated: the General Electric Company was hired to study and to test-launch V-2s. Between 1946 and 1951, sixty-seven captured German V-2s were flown experimentally, mainly at the White Sands Proving Grounds in New Mexico. The program included the development of a number of two-stage vehicles called the Bumper series, eight of which were launched and one of which, on February 24, 1949, penetrated space to set a long-standing world altitude record of 244 miles. On July 24, 1950, Bumper 8 became the first rocket launched from Cape Canaveral, Florida. Under Project Hermes, which continued until 1954, General Electric developed a variety of missiles: the Hermes A-1, a smaller (13,500-pound-thrust, 38-mile-range) finned version of the V-2; the Hermes A-2 (later redesigned as the RV-A-10), America's first large-scale solid-fuel vehicle; the Hermes B, designed for flight-testing a ramjet in the nose of a V-2 (a ramjet is a propulsion engine consisting of a tube rammed through the air at supersonic speed, the air then being burned with injected fuel such as gasoline); the Hermes C, a three-stage rocket glider which never materialized; and the Hermes A-3, an experimental 150-mile-range tactical missile that was a major advance in the effort to develop a reliable engine and a radio-inertial guidance system. Moreover, in 1949 North American Aviation, Incorporated, made copies of V-2 engines which played an unrecognized but enormous role in the subsequent development of America's next generation of large liquid-fuel rocket engines.

V-2s also made it possible for the United States to explore the upper atmosphere and come closer to penetrating space. Besides giving the Americans an education in the fundamentals of large-scale liquid-fuel rocketry, the V-2s were modified to carry a variety of scientific instruments for recording atmospheric composition, pressure, and density;

wind speeds; temperatures; cosmic ray and solar radiation; and other upper-atmospheric phenomena. They were also able to take the first high-altitude photos of Earth. All these experiments were the beginnings of space science. (Even the high-altitude photos were of use. Besides providing new perspectives of Earth, they pointed the way to greater aerial reconnaissance possibilities and appear to have induced the Weather Bureau's Harry Wexler to make the first proposals for weather satellites.)

Another field that was born from V-2 technology was U.S. space biology. On December 17, 1946, a V-2 carried its first biological payload—a package of fungus spores—in order to determine the effects of cosmic rays at high altitude. (The rocket reached a height of 116 miles, a record for single-stage rockets and for the Hermes series as a whole.) Similar cosmic ray exposure experiments were attempted with fruit flies and corn seeds, also in 1946. Not until three years later were higher life forms—four Rhesus monkeys named Albert—permitted to fly by rocket. The experiments were sponsored by the Aero Medical Laboratory at the Wright Air Development Center in Dayton, Ohio. James P. Henry, head of the center's Acceleration Unit, supervised the design of pressurized capsules which were the precursors of the manned cabins used more than two decades later. On June 11, 1948, nine-pound Albert I was anesthetized and prepared for flight, but he perished due to breathing difficulties even before lift-off. On June 14, 1949, Albert II was placed in a better-designed capsule. Telemetry recorded his heartbeats up to the rocket's peak altitude of eighty-three miles, but the parachute was defective and the rocket crash-landed, making Albert II the first space martyr. Alberts III and IV went up September 16 and December 8, 1949, respectively, but one rocket exploded three miles up and the other crashed as the result of a failed parachute. The final V-2 biological payload, launched August 31, 1950, was a mouse, whose reaction to acceleration and weightlessness was recorded by movie camera. Again the parachute failed, but the camera survived.

Now uses for the V-2 rocket in space came from other quarters, too. In February 1945, in the "Letters to the Editor" column of the British magazine *Wireless World,* science fiction writer Arthur C. Clarke suggested a communications satellite that would be launched by a V-2; this was to broadcast "information as long as the batteries lasted," or, indefinitely, by "photo-electric instruments." But in the magazine's

October 1945 issue, in a full article, Clarke proposed a more elaborate plan: he envisioned global radio coverage via three equidistant, geosynchronous (revolving once every twenty-four hours) communication satellites, to be launched by V-2s. (Geosynchronous Earth orbits, or GEOs, are thus sometimes called Clarke orbits, though he was not the first to describe them. In 1897, in his novel *Auf zwei Planeten,* Kurd Lasswitz had described a gravity-nullifying space station that maintained a fixed position about 4,000 miles above the North Pole. But a year later American author Mary Platt Parmele described a true orbiting geosynchronous manned station in her novel *Ariel.* Her fictional station traveled around the world once every twenty-four hours at an altitude of 400,000 miles.)

In 1946, H. E. Ross and R. A. Smith of the British Interplanetary Society conceived a V-2 adapted for an experimental manned ballistic flight. Called Megaroc, the project would have worked much like the later U.S. Project Mercury suborbital manned flights, but it was rejected by the Ministry of Supply, apparently for economic reasons. An American idea was the famous Project RAND satellite proposal, released in May 1946. Based on German experience with the V-2, the RAND "Experimental World-Circling Spaceship" was considered for a variety of potential applications, from weather forecasting and communications to a gravity-free biological testing platform. Unfortunately the RAND vehicle met with indifference from officials. Nonetheless, it was abundantly clear that space rockets had practical applications and that the sword could indeed be turned into a plowshare.

■ RUSSIAN DEVELOPMENTS

In January 1945, as the Russian Army rolled toward Peenemünde from the east, von Braun and the core of his rocket team voted to cast their lot with the Americans. Lest equipment and papers fall to the Soviets, everything was removed from Peenemünde and other rocket facilities. Von Braun then engineered a transfer of almost all Peenemünde's personnel to Bleicherode, in the Harz Mountains. A deserted Peenemünde was seized by the Russians on May 5. But the Russians were as determined as the Americans to acquire rockets and scientists, sending their own missions to grab what they could. The first of these arrived at Peenemünde in June. Heading this group was Grigory A. Tokaty, an aerodynamicist who had taken a keen interest in rocketry ever since he had met Tsiolkovsky in 1933. Also visiting Peenemünde in 1945 were two figures from the past: Valentin Glushko, who had been a member of the Gas Dynamics Lab in the 1920s and 1930s, and Sergei Korolev, of GIRD, who had been swallowed up in the terrible purges of 1937–1938. After a year at hard labor, Korolev had been shifted to a *sharashka* (a special prison for scientists and engineers) and was still nominally a prisoner. But at the close of the war Russia desperately needed his expertise, and he was now serving under heavy guard as Glushko's deputy.

Beginning in 1945, Korolev was given increasingly important roles in Soviet rocketry. He was responsible for refurbishing the underground V-2 production plant at Niedersachswerfen, near Nordhausen, while Glushko was assigned the task of managing test firings of V-2 engines at Lehesten. Korolev was frustrated, however, at not being able to properly witness a demonstration of the last of the few British-captured V-2s, which took place on October 14, 1945, at Altenwalde, near Cuxhaven. (British efforts to examine and experimentally test-launch captured V-2 rockets, with the aid of German technicians, were known as Operation Backfire.) The British had extended invitations to only three Soviet observers: Glushko, Yuri Pobedonostsev (another veteran of GIRD), and General V. L. Sokolov. Korolev had no credentials and could only see the flight outside the barbed wire compound. Still, the Russians had many other opportunities to assimilate V-2 technology.

In June of that year, a day before the arrival of the Russians, members of the U.S. Army's Special Mission V-2 had paid a final visit to Nordhausen to evacuate remaining German rocket scientists. Lacking the necessary authority, they had refrained from blowing up the facility. Shortly afterward, a Soviet Army "reparations" team, scouring Nordhausen for the Ministry of Building Materials, discovered a supply of steam shovels that were desperately needed in Russia. These were parked at the mouth of a large tunnel. Curious, the Russians drove into the tunnel and made far more important finds: V-2 rocket assemblies, tools, plans, intricate radio-directional equipment, and storehouses filled with spare parts. These more precious finds were soon packed in crates and labeled with their destination in the USSR: "To the NKVD [Secret Police] of the Volga-Don." Huge test towers at Peenemünde, which could not be blown up by the evacuating Germans, likewise fell to the Russians.

Moreover, some of the scientists at Peenemünde preferred to work for the Russians rather than for the Americans. Guenther Rosenplanter, an engineer who had developed V-2 steering systems, approached the Soviets in July 1945. Under his direction, Institut Rabe ("Rabe" standing for Raketenbau und Entwicklung, or Rocket Manufacture and Development) was formed to exploit V-2 technology for the Soviets. Rosenplanter was a good organizer but not an outstanding researcher. The Russians then made contact with Helmut Gröttrup, former assistant to Ernst Steinhoff, director of V-2 Guidance and Control. That August or September, Gröttrup suddenly moved to the Soviet zone and became Rabe's new technical chief. Above him was Korolev, the deputy director of the institute. By offering high wages and many privileges, the Soviets induced more Germans to join Rabe, including 200 lower-echelon workers from Peenemünde. The institute's staff thus grew from 30 to 5,000, and by September 1946 was producing flightworthy V-2s. Russian scientists, too, strove to make improvements in the V-2; Korolev, for one, made a number of design changes.

Several important factors now became evident. First, as the Soviet Air Force commander-in-chief told Colonel Tokaty, "Our V-2 rockets do not satisfy our long-term needs . . . What we really need are long-range, reliable rockets capable of hitting . . . the American continent." Second, in practical rocketry, especially in the mass production of large rockets, the Soviets lagged behind the Germans. Third, the Soviets themselves had their own talented rocket people with original ideas. Hence, beginning in the fall of 1946, the Soviets turned their attention to several long-range rocket concepts (they briefly considered Sänger's idea for an Antipodal Bomber, but soon abandoned it as being too grandiose), and they sought to import as much basic large-scale rocket technology as they could into the USSR. Consequently, on October 22, 1946, Gröttrup and a number of other former Peenemünde technicians were suddenly deported to the Soviet Union. There, they were widely dispersed: most of the propulsion experts went to Gorodomlya Island, on Lake Seliger; guidance people to Monino; V-2 developers to Podlipki (later, Kalingrad), and so on. At Podlipki, Pobedonostsev was chief engineer, and Korolev the chief designer. Glushko directed a team of Germans at Khimki.

By 1947 advanced V-2s were being produced at Podlipki, in a converted aircraft factory resembling the Mittelwerk of wartime Germany.

The first German-Soviet V-2, known in the West as the R-10, was variously called the R-1, the 1R, or the Pobeda (Victory) by the Soviets. It was almost identical in appearance and size to the V-2: 46.8 feet long and 5.4 feet in diameter. But its take-off weight was 40,590 pounds, 12,420 pounds heavier than the V-2, and its range was 570 miles. Its RD-100 engine, designed by a team under Glushko, resembled that of a V-2 and likewise ran on a mixture of lox, alcohol, and water, producing 70,400 pounds of thrust (compared with about 56,000 for the V-2). However, the R-1 had a separable warhead (the V-2's was fixed) and a "monocoque" structure in which the tanks were part of the body. On September 30, 1947 (or October 10, 1948, according to which Soviet source one is consulting), the R-1 flew for the first time; the project was directed by Pobedonostsev and Korolev, and the site was the new test range at Kapustin Yar, near Stalingrad (now Volgograd). Ten other flight tests followed. About this time, Korolev drew up plans for very long-range missiles and presented his ideas to Soviet leaders, including Stalin.

Soviet researchers were not unmindful of the scientific value of such powerful rockets. Like the Americans at White Sands, they too began intensive investigations of the upper atmosphere, though the program was always highly secret, since their engines, derived from the V-2, were closely linked to the development of an ICBM (Intercontinental Ballistic Missile). This first family of "geophysical" or "academic" rockets, as Korolev called them, were based on the R-1 and included such vehicles as the R-1A, the R-1B, the R-1V, the R-1D, and the R-1E. (Alternative designations of these models were the V-1A, the V-1B, and so on.) On May 24, 1949, the R-1A reached an altitude of 68 miles carrying two side containers of instruments (a total of 375 pounds) which were ejected at the top of the trajectory and recovered by parachute. The R-1B, the R-1V, and the R-1D carried greater payloads; they had long, pointed nose cones and bulging instrument containers on each side. The biological phase of this program, also begun in 1949, was quite advanced. In addition to recording data on solar spectra and on the ionosphere, these rockets carried and filmed dogs (sometimes two in a single vehicle) throughout the flights; rabbits, hamsters, and other animals were also carried.

In 1949, too, Gröttrup's team was asked to leap ahead and design a regeneratively cooled engine that could hurl a 6,600-pound warhead 1,800 miles. This vehicle, designated the R-14, was to run on a mixture

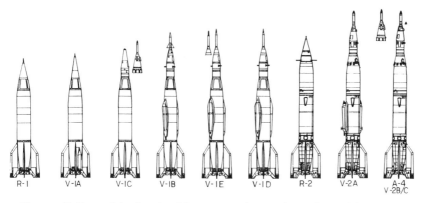

Figure 7. Part of the family of Soviet sounding rockets, derived from the V-2. Based on Soviet drawings.

of lox and alcohol and yield a thrust of 220,000 pounds. Such a missile was an entirely new departure from the V-2 configuration. Its design occupied the Germans until the remainder of their stay in the USSR, but the rocket was never built.

In the meantime, second-generation V-2 derivatives were produced. The R-2 and R-2A (the latter developed in 1954) used the "stretched" V-2 engine, the RD-101, with a thrust of about 80,000 pounds. (The engine used 92 percent ethyl alcohol and 8 percent water, whereas the RD-100 used 75 percent ethyl alcohol.) The R-2A was 57.7 feet long and weighed 45,000 pounds. The use of aluminum instead of steel lightened the rocket, which had double the range of the R-1 series. A "strategic" version (that is, one designed to carry a warhead) is known. The next family of geophysical and "intracontinental" rockets consisted of the R-5s, which used the RD-103 engine; these were developed starting in 1956 and were in service by 1958. For its geophysical applications, the RD-103, which still looked like a V-2 engine, ran on a mixture of 92 percent ethyl alcohol and 8 percent water, yielding a maximum thrust of 97,120 pounds. A variation of the RD-103 may have run on lox and kerosene, and may have powered the intracontinental rocket known to the West as the medium-range SS-3, or Shyster. If the RD-103 did use these higher-performance propellants, it was the first Soviet engine to do so. It would have marked a major advance in propulsion and would have served as the foundation of all future Soviet launch vehicles, including the Sputnik rockets.

The Soviets thus continued with V-2 derivative engines and vehicles until quite late, and the R-5 series was used up to the late 1970s. In the meantime, however, other important Soviet rocketry developments proceeded apace. Earlier, when the Russians had absorbed all they could from Gröttrup and his men, they asked the Germans to return home. Korolev, after a second period of internment in a *sharashka*, was "rehabilitated" following the death of Stalin. In the early 1950s he introduced a wide-eyed Premier Khrushchev to the possibilities of far larger rockets. The stage was now set for a new era of missiles and space exploration.

4

ROCKETS ENTER THE SPACE AGE

R obert Goddard was the first to think of using rockets to gather data in the upper atmosphere at altitudes of greater than twenty miles—that is, beyond the range of unmanned balloons. Such vehicles, which were not capable of exploring space, became known as sounding rockets.

■ SOUNDING ROCKETS

Goddard's proposal for the development of these vehicles gained him Smithsonian support. *A Method of Reaching Extreme Altitudes* was forwarded to Charles G. Abbot, director of the Smithsonian Astrophysical Observatory, who recommended funding Goddard's research on important "meteorological problems." For Abbot, these included such things as the composition and temperature of the highest levels of the atmosphere, as well as phenomena of solar physics. In 1929 Goddard met with top officials of the Weather Bureau, the Mount Wilson Observatory and the Carnegie Institution, who made a number of additional proposals: taking solar spectrograms and solar corona photos, collecting air samples, studying cosmic rays, determining the ion content of the upper atmosphere, and measuring magnetic intensities. Optimistically, the Carnegie Institution granted $5,000 for the anticipated research. But Goddard was wrapped up in perfecting the engine. With the exception of the barograph on his ill-fated 1929 rocket, he never found the opportunity to adapt his rocket to upper-atmospheric and other investigations.

In 1942, in wartime Germany, von Braun awarded a contract to the Forschungsgestelle für Physik der Stratosphäre (Research Foundation for the Physics of the Stratosphere) to develop instruments that would be carried by the V-2 for high-altitude measurements. These included a quartz barograph, a recording thermometer, an ultraviolet spectrograph, and an air sampler. Von Braun justified the purely scientific aspects of this research by stressing military needs; these included steps toward solving problems of trajectory and of frictional heating, as well as measuring effects of the rocket's exhaust upon radio communications and other variables. Pioneer solar physicist Erich Regener was assigned to Peenemünde to develop the instrumentation, but a British bombing raid on August 17, 1943, curtailed this research: part of the ultraviolet spectrograph was apparently destroyed, and the imminent capture of Peenemünde by the Russians in the spring of 1945 ended the work altogether. Not until the postwar V-2 flights at White Sands and Kapustin Yar were true sounding rockets finally established.

Earlier, in 1936, physics students Frank Malina and William Bollay, with experimenters John W. Parsons and Edward S. Forman, had started the GALCIT (Guggenheim Aeronautical Laboratory, California Institute of Technology) Rocket-Research Project to design a solid- or liquid-fuel high-altitude sounding rocket; others joined later. Caltech's president, Robert A. Millikan, was interested in these vehicles for cosmic ray research and greatly encouraged the group, which was directed by the celebrated aerodynamicist Theodore von Kármán. The project set out to investigate the requirements and thermodynamics of sounding rockets, to conduct modest experiments on gaseous and liquid propulsion, and to undertake some solid-fuel studies. By 1939 its work was so promising that the Army Air Corps (now the U.S. Air Force) provided it with financial support for research on JATOs. During the war, the research expanded and the project achieved a number of significant milestones: the application of America's first successful solid-fuel JATO (1941); the founding of Aerojet (1942), which was to become a leading manufacturer of rockets; the establishment of the Jet Propulsion Laboratory (1943), today part of NASA; and the creation of the WAC-Corporal rocket, America's first operational sounding vehicle (1945).

World developments "dictated" the group's work on the military applications of rocket propulsion, as Malina later wrote, but he and his colleagues never lost sight of their initial goal. Indeed, the WAC-

Corporal started as a smaller test version of the Jet Propulsion Laboratory's Corporal missile, conceived in 1944. Then Malina "suddenly realized" that a sounding rocket was "within reach." On January 16, 1945, he submitted to Army Ordnance a proposal for a rocket that would carry a recoverable 25-pound payload to an altitude of 100,000 feet. The project which was distinguished by its elegant simplicity and which promised to be valuable for military needs, was approved by the Army and set a standard for future liquid-fuel sounding rockets.

Reliability and low cost were achieved by pressure-feeding the propellants into the combustion chamber rather than using pumps. The propellants (red fuming nitric acid, or RFNA, and aniline) were hypergolic and needed no ignition system. The fuels were also "storable," so the rocket did not have to be fueled just prior to launch; lox, in contrast, had to be used immediately, since it evaporated. (Parsons thought of using the acid as an oxidizer in 1940, but Sänger had suggested it in 1933, and the German Helmut von Zborowski had begun experimenting with it in 1939. Malina is credited with first using aniline as a fuel.) RFNA is corrosive, but the tanks were built of corrosion-resistant, heat-treated steel. The WAC-Corporal attained high initial acceleration (and consequent stability) by means of a powerful, short-duration, solid-fuel booster, a modified Tiny Tim antiaircraft rocket with the warhead removed. This also did away with the need for complicated gyroscopes and servos. Direction was controlled by three stabilizing fins and a 100-foot launch tower. The WAC-Corporal was 16 feet long (not counting its 8-foot booster) and one foot in diameter, and weighed 665 pounds loaded. Its Aerojet regeneratively cooled motor produced 1,500 pounds of thrust for 45 seconds. The booster, which weighed 759 pounds, produced 50,000 pounds for 0.6 seconds.

The WAC-Corporal first flew at White Sands on October 11, 1945. Radar tracked it to an altitude of nearly forty-five miles, twice as high as originally specified. Other successful tests followed, but when captured V-2s appeared in the New Mexico desert, WAC-Corporals were no longer needed, though some served as second stages in the Bumper series and became the first vehicles to penetrate deep space.

As the stock of V-2s dwindled to about twenty-five, the Applied Physics Laboratory of Johns Hopkins University requested Aerojet to develop a successor to the WAC-Corporal for carrying 150 pounds of instruments to altitudes of fifty miles and more. The result was Aerobee, first launched on November 24, 1947. This rocket, which ran

Figure 8. Top: Frank Malina beside his WAC-Corporal. Bottom: Aerobee-Hi at White Sands, New Mexico.

on RFNA and aniline and also had a solid-fuel booster, enjoyed thirty-eight years of success: the last of the series (Aerobee 1,058) was launched on January 17, 1985. During this period a dozen models were manufactured: the standard Aerobee with 2,600 pounds of thrust, the Aerobee-Hi, the Aerobee 75 (or Aerobee Hawk), the Aerobee 100 (or Aerobee Junior), the 150, the 170, the 170A, the 200, the 200A, the 300, the 300A, and the 350. The last was powered by four clustered motors yielding a total thrust of 16,400 pounds and could reach altitudes of up to 290 miles. Aerobees have lofted every conceivable high-altitude payload and have made significant discoveries, most prominently during International Geophysical year (1957–1958), a period of intense solar activity in which many nations cooperated to probe the inosphere. Aerobees have lifted mice and monkeys to altitudes of up to thirty-eight miles (1951–1952); taken the first color photographic image of Earth from space, using a specially designed camera with filters, an important contribution toward later meterological satellites (1954); mapped the stars using ultraviolet light (1960); collected micrometeroids (1961); taken solar coronagraphs (1966); detected X-ray stars and perhaps the first neutron star (1970; and mapped airglow (an aurora-like phenomenon in the upper atmosphere), photon densities, and electron densities in the upper atmosphere. In addition to accomplishing these scientific breakthroughs, in 1957 the Aerobee pioneered the use of an invaluable technological tool known as the Sun Seeker. Above a certain level of the atmosphere (about 15 miles), rocket fins are no longer effective in stabilizing a sounding rocket. At that point, the rocket oscillates considerably, making measurements extremely difficult to take. Mounted on the Aerobee's nose, the Sun Seeker is an armlike device that swings freely and has photoelectric cells on its tip. The cells keep the arm pointed in the direction of the sun by hydraulic means, despite the gyrations of the rocket. Thus, spectographs or other instruments mounted on the Sun Seeker are provided with a stable platform to make their measurements.

Meanwhile, a far bigger sounding rocket than the Aerobee made its own technological and scientific gains. In 1946 the U.S. Naval Research Laboratory realized that a heavy-lift rocket was needed to replace the V-2. This led to the development of the Neptune, later renamed the Viking. That summer, the Glenn L. Martin Company was hired to make ten rockets; four more were subsequently ordered. Reaction Motors, Incorporated, made the engines. All Vikings differed slightly, but there were two basic types: the Viking 1-7 and the Viking 8-14, which were 32 inches and 45 inches in diameter, respectively. Average height was 46 feet. Loaded weight ranged from 9,650 pounds to 15,030 pounds, while average thrust was 20,600 pounds. The Viking was America's first large-scale rocket; like the V-2, its XLR-10 engine burned a mixture of lox and alcohol, which was fed by two turbine pumps activated by the decomposition of hydrogen peroxide. However, the XLR-10 was regeneratively cooled and was gimbaled by linkage to an autopilot. Gimbaling—meaning that the thrust chamber could be tilted along four different axes for precise steering throughout the rocket's flight—marked a significant advance over jet vanes, which were complex and heavy. (Partial gimbaling had been used by Goddard

Figure 9. The Viking 4 being launched from the deck of the USS *Norton Sound,* May 10, 1950.

beginning in 1937, had been suggested for the V-2 but never tried, and would be fully incorporated for the first time in the short-lived U.S. MX-774 missile in 1948.) The Viking's stability, which was critically important for taking scientific measurements in the upper atmosphere, was also achieved by two movable fin tabs and by small hydrogen peroxide thrust jets placed at various points around the rocket. Viking 1 was launched May 3, 1949. The twelve vehicles in the series were successful with the exception of Viking 8, which broke away during a static test and was destroyed. In 1954 Viking 11 set a single-stage altitude record of 158 miles. All but three of the vehicles were fired from White Sands (Viking 4 took off from the deck of the USS *Norton Sound,* near Jarvis Island in the Pacific; and two additional Viking rockets, used not in the Viking program but as test vehicles for the Project Vanguard satellite project, were launched in 1956 and 1957 from Cape Canaveral). Valuable ionospheric and other data were obtained through the program. But the Viking's exorbitant cost of $400,000 per launch, to say nothing of its ground-support requirements, doomed large liquid-fuel sounding rockets to extinction and inevitably led to the development of smaller, cheaper solid-fuel systems. In addition, the introduction in the early 1950s of "composite" solid fuels (based on polysulfides and perchlorates) permitted the construction of larger and more powerful solid-fuel rockets than did double-base propellants.

The ancestor of the solid-fuel sounding rockets was the Deacon, a bargain at $4,000 per vehicle, and capable of carrying payloads of up to 40 pounds. Development of the Deacon began in 1945. At first, it was a versatile tool used by the National Advisory Committee for Aeronautics (NACA) to boost model aircraft to supersonic speeds for aerodynamic testing. The Deacon used a double-base fuel and produced about 6,000 pounds of thrust for 3.2 seconds. In 1954 NACA outfitted Deacons with Nike-Ajax antiaircraft rocket boosters (also costing $4,000 each), in order to achieve higher Mach numbers; thus was created the Nike-Deacon, or Dan. Dans were soon being used to carry 50-pound payloads to altitudes up to 69 miles, for measuring upper-air densities and temperatures. Since then, the Dan's powerful boosters, yielding 50,000 pounds of thrust for 3 seconds, have been mated to a wide variety of other solid-fuel rockets, begetting an extended family of effective, low-cost, very reliable sounding vehicles. Among these are the Nike-Cajun, the Nike-Asp, the Nike-Recruit, the

Nike-Dart, the Nike-Javelin, the Nike-Tomahawk, the Nike-Apache, the Nike-Orion, the Nike-Hawk, the Nike-Iroquois, the Nike-Malemute, and even the Nike-Nike. The same "off-the-shelf" (surplus) stacking principle has been used to create the big (90,000-pound thrust) Honest John solid-fuel missile motor, which has made possible the construction of higher-performance sounding vehicles such as the Honest John–Nike, routinely lofting payloads of up to 500 pounds to altitudes of over 300 miles. Single-stage rockets, from the Arcas to the Wasp, also abound, contributing, along with foreign sounding rockets, to the world-wide data-sharing Committee on Space Research, founded during the International Geophysical Year (1958) by the International Council of Scientific Unions. Today the United States participates in numerous sounding-rocket programs in cooperation with other nations.

Another ingenious offspring of the sounding rocket was the "rockoon," or balloon-launched rocket—a cheap way of reaching higher altitudes. Goddard (1907), von Hoefft (1928), and others had thought of this as a means for launching space vehicles, in order to avoid lower-atmosphere friction. But it was physicist James Van Allen, along with Commander M. Lee Lewis of the Office of Naval Research, who applied the concept to sounding rockets. The initial firing used a plastic balloon and launched from the Coast Guard Cutter *Eastwind* off Greenland in 1952. A Deacon suspended from a balloon was ignited by radio command and reached an altitude of 50 miles. Rockoons were soon well established: 86 of them probed the upper atmosphere during the International Geophysical Year, when a total of 210 rockets were fired in the United States.

In that year, the Soviets were still using their advanced V-2 derivatives, from the V-2A to the V-5B. These were also designated A-1 to A-4, signifying that they carried payloads for the USSR Academy of Sciences. In addition to inospheric research, these vehicles continued to carry out biological experiments. In August 1959 one flew to an altitude of 279 miles with a recoverable payload of two dogs, scientific apparatus, and electronic equipment—a total of 3,725 pounds. Between 1948 and 1950 the Soviets developed a "Meterological Rocket," called the Meteo, or the MR-1. This ran on a mixture of nitric acid and kerosene, had solid-fuel boosters, and reached a peak altitude of 62 miles. It was extensively used during the International Geophysical Year, when the USSR fired a total of 175 rockets. In the late 1970s more advanced Meteos, such as the solid-fuel MR-12 and the two-stage solid-fuel MR-100, were fired from the *Akademik Korolev* and other research ships off NASA's Wallops Flight Center in Virginia. These launchings were part of an unusual joint U.S.-USSR program to gather stratospheric and mesospheric data. More spectacular were the ten vehicles—each 75 feet tall—in the Vertikal series, launched between 1970 and 1981. Vertikals, which were modified nuclear-armed IRBMs (Intermediate-Range Ballistic Missiles), lifted payloads of 2,000 pounds up to altitudes of over 900 miles for the USSR's Intercosmos program, in which Soviet-bloc countries conducted the experiments.

France, too, pioneered in the development of sounding rockets. Like the United States and the USSR, France acquired V-2 scientists from Germany. After World War II, Wolfgang Pilz and other researchers from Peenemünde began building liquid-fuel missiles for the French. Subsequently, in March 1949, they were asked to create a sounding vehicle, which became known as the Véronique. This rocket resembled the V-2 but burned white fuming nitric acid mixed with diesel oil, turpentine, or alcohol, to produce a thrust of 8,820 pounds. It was first launched in 1950, with later tests in Hamaguir, Algeria. Eventually, the Véronique became France's Aerobee; it was used extensively for atmospheric and astronomical experiments until the 1970s. France also built many other vehicles, such as the Vesta, the Bélier, the Centaure, the Dauphin, the Dragon, the Eridan, and the "diamond series" (with names like Rubis, Topaz, and Emeraud).

Similarly, at the end of the war England obtained the services of Walter J. H. Riedel, von Braun's deputy, who, after a brief stint at Britain's Royal Aircraft Establishment, worked at the Rocket Propulsion Establishment in Westcott until his death in 1968. The English, however, were perennially short of funds and were unable to expand their Project Backfire, based on captured V-2 technology. Rather, they made use of Riedel's expertise mainly to develop the RTV-1 (Research Test Vehicle 1), which ran on lox and alcohol and which led to other relatively small missiles. Not until 1956 did they create their first sounding rocket, a vehicle named the Skylark. Since the Skylark's solid-fuel motor (the Raven) was developed at the Rocket Propulsion Establishment and since Riedel did some solid-fuel research there, it is likely that he was involved in later Raven improvements.

The first Raven produced a nominal thrust of 11,500 pounds for 30 seconds, but the Skylark has since undergone many modifications and has acquired boosters and stages. To date, more than 350 Skylarks have been launched; they have collected data in a variety of fields—including astronomy, remote sensing, and microgravity experiments—and have been used in international scientific programs in Australia, Sweden, Spain, Argentina, Norway, and Sardinia. Later vehicles, such as the three-stage Skylark 12, have lifted payloads of 770 pounds to altitudes of over 620 miles. On August 25, 1987, the Skylark celebrated its thirtieth anniversary by carring a West German X-ray telescope and camera to photograph a recently discovered supernova 155,000 light years distant. Besides Skylark, England also has several smaller sounding vehicles: the Petrel, the Fulmar, and the Skua.

Elsewhere, preparations for the International Geophysical Year prompted Japan to take up scientific rocketry. By 1955 Japanese scientists had developed miniature "Pencil" rockets that were 9 to 12 inches long and 0.7 inch in diameter. Diminutive though they were, they provided fundamental data on propellants, rocket dynamics, and tracking. From this very modest start, Japan rapidly graduated to the Kappa series of solid-fuel rockets; by 1958, Kappa 6 was capable of lifting an 11-pound payload to an altitude of 135 miles. The Japanese also tried rockoons. By the mid-1960s, their four-stage Lambda rockets were proving to be as successful as sounding vehicles in other countries and were preparing Japan to enter the ranks of space nations.

Since sounding rockets cost relatively little, many other countries have also sought to participate in meaningful research in the upper atmosphere to the fringes of space. For example, Australia began with its Long Tom rocket, India with its Rohini, Canada with its Black Brant, and Brazil with its Sonda. Since the advent of satellites and shuttles, the need for sounding rockets for probing the lower levels of the stratosphere and ionosphere has remained vital and, indeed, requires global participation. Sounding rockets, which ushered in the Space Age, will therefore continue to be used well into the future.

■ THE FIRST SATELLITE LAUNCHERS

The V-2 sired the artificial, unmanned satellite. Before the V-2, almost nothing had been written about satellites. In the days of Tsiolkovsky and Goddard, researchers concentrated on manned flight. It was also believed that radio waves could not penetrate a certain layer of the upper atmosphere—the so-called Heaviside layer. More important, in those years of bulky radio tubes, there simply was no adequate electronic technology for satellite instrumentation; and telemetry did not yet exist. For these reasons, radio communications studies are conspicuously absent from the early astronautical literature. Beginning in 1945, however—given large-scale rocket technology, wartime advances in electronics, and the development of transistors (first introduced in 1948)—proposals for satellites proliferated. In addition to unlocking the secrets of space beyond the ionosphere (above 200 miles), satellites promised to facilitate worldwide radio communications and to provide instant monitoring of global weather patterns. However, it took a decade for the United States and the USSR to design and implement programs so that they could launch satellites during the International Geophysical Year. In January 1955 Radio Moscow had announced a prospective satellite launching but this and subsequent similar announcements were not considered seriously. The United States had long taken its technological superiority for granted; besides, satellite and space station proposals were at that time quite common, and most of them were considered pie-in-the-sky fantasies.

But on July 29, 1955, the White House announced that the United States would launch "small earth-circling satellites" as part of its contribution to the International Geophysical Year. In the following months, the Department of Defense, the National Science Foundation, and other agencies considered several satellite schemes. On September 9 it was announced that Project Vanguard, developed by the Naval Research Laboratory, had been chosen. As planned, the Vanguard was a nonmilitary vehicle consisting of three stages: a modified Viking stage with a General Electric Hermes A-3B engine; an upgraded Aerobee-Hi stage, either pressurized or pumped, which ran on a mixture of unsymmetrical dimethylhydrazine (UDMH) and white inhibited fuming nitric acid (WIFNA), which were storable propellants; and a solid-fuel third stage. When actually built, the first stage used a gimbaled GE X-405 engine (more complex than the Hermes), which produced 27,000 pounds of thrust and burned a mixture of lox and kerosene. Like the V-2, the X-405 was regeneratively cooled, and was fed by turbine pumps initiated by a peroxide steam generator. Its rocket casing, 45 inches in diameter, was built by the Martin Company using Viking sounding-rocket tooling. The second stage used a helium-pressurized Aerojet AJ-10 engine, which produced 7,500 pounds of thrust. The third stage was a Grand Central Rocket Company 33-KS-2800 steel unit, which generated 2,800 pounds of thrust for 33 seconds; this was later replaced in Vanguard III by a lighter engine developed by the Allegheny Ballistics Lab—the innovative X-248 Altair, which was made of glass-reinforced plastic and yielded a thrust of 3,100 pounds. (The Altair's construction was considered an important advance in solid-fuel rocket technology.) Overall, the Vanguard vehicle was 72 feet tall and weighed 22,600 pounds.

In December 1956 and May 1957, the last two Viking rockets were wheeled out to serve as Vanguard test vehicles (TV-0 and TV-1), so that telemetry and other functions could be evaluated. TV-1 carried a second stage which was a prototype of Vanguard's third, solid-fuel stage. Both tests went flawlessly, the final Viking making a beautiful night ascent. Although the TV-3 was encountering some delays, everything seemed to bode exceptionally well for the program.

Then, on October 4, 1957, the entire world received a shock: the Soviets launched Sputnik 1. The USSR was the first nation in space. Americans everywhere suddenly lost their complacency over their presumed technological superiority. The successful first flight of the TV-3 on October 23 did not soothe matters, since the national trauma was almost immediately compounded: on November 3 the USSR launched the much larger Sputnik 2, carrying Laika, a dog, the first living organism in space. The third shock came on December 6: the TV-3, the first complete Vanguard, exploded in a ball of flame less than one second after lift-off, pathetically dropping its four-pound spherical test satellite.

Now, with national prestige at stake, von Braun and his team at the Army Ballistic Missile Agency received the green light to launch a proven military vehicle, a modified Jupiter-C, carrying a satellite designed by the Jet Propulsion Laboratory. On January 31, 1958, America's first satellite was launched: the Explorer 1, weighing 10.5 pounds. But this was countered on May 15 by the USSR's Sputnik 3, an immense "flying laboratory" of 7,000 pounds. On March 17 America's Vanguard 1 (TV-4) finally made it into space; it was the first satellite to incorporate solar cells and, still in orbit today, is designed to remain so for 1,000 years. Vanguard satellites 2 and 3, launched February 17 and September 18, 1959, brought the program to an end; like Vanguard 1, they are still in use.

In contrast to satellite programs in the United States, those in the

Figure 10. The R-7, the world's first satellite launcher, which lifted the Sputnik 1 to orbit on October 4, 1957.

 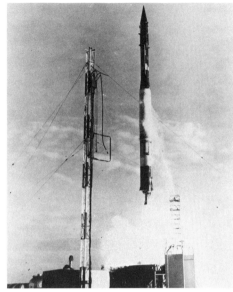

Figure 11. Left: a Jupiter-C preparing to launch the Explorer 1. Right: a Project Vanguard Test Vehicle (TV-4) launching a Vanguard 1 satellite.

Soviet Union have been conducted in secret, well removed from public scrutiny. For a decade, the Soviets released neither pictures nor data of the first Sputnik launchers, refusing to reveal even the name of their rocket designer, Korolev. Until his death in 1966, he was referred to only as the "chief designer." In fact, it was through Korolev's "crazily intensive effort," as one of his Soviet biographers has said, that the Soviet government decided to launch its own satellite in honor of the International Geophysical Year; the official announcement was made July 30, 1955. Korolev's idea was to use the USSR's first ICBM, the R-7, nicknamed Semyorka (meaning "Ol' Number 7") and afterward known in the West as the SS-6, or Sapwood. This vehicle was powered by the RD-107 and the RD-108, the former being the Soviet Union's first native rocket engine. (Significantly, the engine was built by the Gas Dynamics Lab–Experimental Design Office, or GDL–OKB, which was a successor to the original GDL of the 1920s and 1930s.) Designed in 1952 by Korolev and Aleksei Isayev, the RD-107 borrowed many V-2 techniques but differed in two principle ways from earlier

RDs: it burned a mixture of lox and kerosene, and it used clustered combustion chambers with two steering verniers. Isayev, who in the 1940s had worked on rocket fighter planes, may have been the one to first suggest clustering a smaller engine, but Korolev considerably upgraded the idea. In the clustered approach, which characterizes almost all later vehicles in the RD series, four chambers are fed by a common turbo-pump, usually driven by hydrogen peroxide steam (as in the RD-107). The RD-107, delivering 224,910 pounds of thrust, was both regeneratively cooled and film cooled and had bronzed fire walls (resistant to high heat) with bronze swirl-type injectors. The RD-108 was similar but had four verniers, a thrust of 211,680 pounds, a longer burning time, and different starting controls.

The problem in building a nuclear ICBM in those days was that rockets still had relatively low thrust, whereas first-generation thermonuclear weapons were extremely heavy, amounting to warheads of about 7,000 pounds. Korolev's solution was a "cluster of clusters." In this configuration, the R-7's central or "sustainer" engine was a single RD-108; surrounding the rocket were four RD-107s serving as boosters, which were jettisoned after use. Thus, the R-7's total of 32 large and small vernier combustion chambers collectively produced more than 1.1 million pounds of thrust at lift-off. To save weight and to streamline the rocket, the boosters were placed in tapered pods with a fin protruding from each. The overall diameter for the Sputnik version of the R-7 was 33.7 feet, the length was 95.69 feet, and the estimated loaded weight was 588,735 pounds. This pragmatic design, though not very efficient in terms of weight and cost, resulted in what was then the world's largest and most powerful rocket. It served as the basis for much of the Soviet Union's space program from 1957 onward, and was not outclassed until the United States first launched the Saturn IB, yielding 1.6 million pounds of thrust, in 1966.

Hardware for the R-7 was developed between 1954 and 1957. After several failures, the first successful test flight was made August 3, 1957, from the new Baikonur launch complex near Tyura-Tam in Kazakh S.S.R. On August 27 came the first full-range flight: more than 5,000 miles. Six weeks later, on October 4, the 184-pound Prostiayshii Sputnik 1 (Preliminary Satellite 1, or PS-1) was launched. The same vehicle also launched Sputnik 2 (PS-2) and Sputnik 3. In the West, there was a lot of confusion and speculation as to the identity of the first Sputnik launchers. Indeed, until the 1960s Western intelligence sources (using reconnaissance means) simply did not know about the Soviet use of strap-on boosters and Korolev's cluster-of-clusters configuration.

Jupiter-C, launcher of America's Explorer 1, had its own hidden history. In 1945, awed by the V-2, the Air Forces of the U.S. Army asked several firms to examine the possibility of long-range missile development. In July 1946 a contract was signed with North American Aviation, Incorporated, to begin Project Navaho. Originally envisoned as a subsonic missile with a range of 500 miles, the Navaho was later upgraded to a supersonic ramjet weapon with a range of 5,500 miles and liquid-fuel boosters yielding 415,000 pounds of thrust. North America's only previous experience with rockets had consisted of test-firing experimental engines of about 2,000 pounds of thrust in a corner of their parking lot. But by 1949 William Bollay, technical director of North American's Aero Physics Lab and a former member of the 1936 GALCIT Rocket Research Project, had obtained plans and components for V-2 engines from White Sands. These were studied closely by North American's engineers, who even cut apart some of the components to see how they were made. North American then built three copies of the engine and static-fired them. In this way, said rocket engineer George P. Sutton, who was in charge of the tests, "we jumped from the 2,000-pound thrust range right up to the 50,000-pound range." However, these were not precise copies of the original engine. The Americans had no German materials; they used standard American fittings (threads, O-rings, and so on), and the parts were not interchangeable with German ones. But the copies were reasonably close, and their accuracy was enhanced through the advice of experienced German rocket scientists. Most notably, Sutton consulted with von Braun, then at Fort Bliss, Texas; and North American employed Dieter Huzel, formerly in charge of Peenemünde's main test stand, and Walter Riedel, also from Peenemünde (known as "Riedel II," he was unrelated to Walter J. H. Riedel). North American was thereby able to assimilate V-2 technology very rapidly, and to incorporate it into their Navaho boosters. This technology transfer was to have enormous consequences.

Navaho's XLR43 (Experimental Liquid Rocket 43) powerplant was first fired in March 1950; producing 75,000 pounds of thrust, it was America's first large-scale liquid-fuel engine. An upgraded version (XLR71) with an output of 100,000 pounds was tested in November 1952; the 200,000-pound mark was reached in August 1953; and in

January 1956 a three-chambered version (XLR83) delivered 415,000 pounds. Yet although the program was canceled in July 1957 for budgetary reasons, the Navaho fathered the first generation of America's large liquid-fuel rockets that also served as launch vehicles. The Redstone, America's first intermediate-range ballistic missile, conceived in 1951, initially used a modified XLR43 lox-alcohol Navaho engine, a shortcut that saved time and money. Designed by von Braun's team from Peenemünde, the Redstone exhibited many V-2 features, such as exhaust vanes and V-2–type firing table. The Redstone in turn, gave rise to the Jupiter IRBM, which used a modified Redstone engine that burned lox and kerosene fuel and that produced a thrust of 150,000 pounds. It had regenerative cooling and a gimbaled, bell-shaped chamber. Jupiter's inertial guidance system and ground-support equipment were also patterned on those of the Redstone. Another offspring of the Redstone was the Jupiter-C (also called the Juno I), a four-stage rocket that launched the Explorer 1. The Juno II, a Jupiter with upper staging from the Juno I, launched america's first lunar probe, the Pioneer 3, as well as the Pioneer 4 and Explorers 7, 8, and 11. Incidentally, early in the Space Age, both the Americans and the Soviets used Roman numerals to designate some of their spacecraft and vehicles, but as these projects proliferated it became more practical to give the designations in Arabic numerals.

The Thor IRBM, developed at the same time as the Jupiter, used a similar Rocketdyne engine with 150,000 pounds of thrust. (In 1955 North American's rocketry division had been named Rocketdyne.) And in the years from 1954 to 1957, there were numerous other breakthroughs: smaller nuclear payloads of 1,500 to 3,000 pounds; H. Julian Allen's "blunt-nose principle" for slowing down reentry speeds; heat-absorbing ablation nose cones; and more precise all-inertial (automatic, nonradio) guidance systems. As a result, the Atlas, America's first ICBM, which had been delayed by technical problems, began to make good progress. To speed its development, three engines adapted from the aging Navaho were incorporated in its design, two "outboard" Rocketdynes as dispensable boosters and one inboard Rocketdyne (with 60,000 pounds of thrust) as the sustainer. Total liftoff thrust was 330,000 pounds. Atlas represented a considerable advance toward future generations of manned and unmanned launch vehicles. Outstanding contributions were designer Karel J. Bossart's combined use of boosters and sustainer arranged in a row, so as to eliminate complicated staging; thin-skinned, internally pressurized fuel tanks (an idea originating with Oberth in 1929 but directly adapted from Atlas's predecessor, the MX-774, of 1946); use of the same fuel tanks in boosters and sustainer; and stainless steel monocoque construction, which made possible a lightweight, heavy-lift vehicle. Of a different configuration, though using compatible components, was the simpler, lighter, two-stage Titan ICBM, developed beginning in 1955 to further close the "missile gap." The first stage consisted of two Aerojet engines yielding a total thrust of 300,000 pounds; the second, a single 80,000-pound motor of the same model. The Atlas's engines would have been compatible with the Titan's.

The Jupiter-C (or Jupiter Composite Reentry Test Vehicle) was a pioneering vehicle in the development of upper stages. It was an upgraded Redstone with two extra stages, so that ablation nose cone materials could be tested for the Jupiter. These stages propelled the Redstone toward a reentry trajectory. Stage two was composed of eleven solid-fuel Baby Sergeant rockets, which were arranged in a ring 37.8 inches in diameter and which produced a combined thrust of 17,600 pounds for 6 seconds. Three identical rockets inside the ring made up the third stage, producing 4,800 pounds. The whole was set in an electrically rotated tub to provide ballistic stability. The first Jupiter-C, launched in September 1956, attained an altitude of 682 miles and a horizontal distance of 3,300 miles. Von Braun knew that with the addition of a single fourth-stage rocket, a Baby Sergeant with a small scientific payload attached could easily be set in orbit around the Earth. "All efforts to extend our technology into space were unsuccessful," he later wrote, until the shock of Sputnik 1. Only then was he given permission to proceed, and the result was the Explorer 1.

Other major achievements for the Redstone, whose lineage traced back to the V-2 and the Navaho, were America's first Project Mercury manned flights. Project Mercury was approved within a week after the National Aeronautics and Space Administration (NASA) was formed on October 1, 1958. The Redstones, called "old reliables," were soon "man-rated" (deemed safe for manned flight). As with the Jupiter-C, the fuel-tank section was lengthened by six feet for longer burning time. An adapter was devised to link a Redstone engine to the one-man Mercury spacecraft; this was topped with a heat shield and escape tower. (The tower was a triangular mast attached to the top of the spacecraft; the end of the mast was fitted with a long solid-fuel rocket having three canted nozzles. In the event of a mishap early in a mis-

sion, the escape system would instantly boost the spacecraft away from the launch vehicle. On flights that proceeded normally, the tower was automatically jettisoned from the vehicle after a safe altitude had been reached and the boost engine had cut off.) An autopilot control system was also installed in the spacecraft, as well as an abort-sensing system. On May 5, 1961, Alan Shepard, Jr., was launched aboard the MR-3 (Mercury-Redstone 3) and became America's first man in space; on July 21 Virgil I. Grissom flew the MR-4. Subsequent Mercury missions used the Atlas. The two-man Project Gemini capsules, which were heavier, used a modified second-generation Titan 2 ICBM.

The building-block concept for achieving both manned and unmanned spaceflights was mastered by the Soviets as well. Following the launch of Sputnik 3, the R-7 ICBM was further modified by the addition of an upper-stage engine that ran on lox and kerosene. This engine was produced by the Experimental Design Bureau (OKB), headed by Semyon Kosberg (a former airplane-engine designer) and thus known also as the Kosberg Bureau. The modifications increased the R-7's payload capacity: it could lift more than 12,000 pounds into Earth orbit, and enable about 880 pounds to escape Earth's gravity completely.

Naturally, the Soviets took every advantage of this early weight-lifting lead. In January 1959 they sent Luna 1 toward the moon. This probe missed its target and went into solar orbit instead. But Luna 2, launched in September, successfully landed on the moon. From May 1960 to March 1961, the Soviets used a more powerful upper stage, also designed by Kosberg. This enabled them to launch a series of five Korabl Sputniks (manned satellites), which were actually precursors of manned spacecraft, since they carried either dogs or mannequins and were designed for recovery. Here, the precision liquid-fuel retro rockets built by Isayev were critically important, although they did not always function properly. (A retro rocket is a small rocket engine that uses either liquid or solid fuel and that is mounted on a larger rocket or spacecraft. Its nozzle faces in the direction opposite to the thrust of the carrier vehicle's sustainer engine. It therefore provides reverse thrust, which is used for decelerating the carrier vehicle during orbit descents, landings, and other maneuvers.)

On April 12, 1961, the 125-foot-tall R-7, renamed the Vostok (East), with four verniers and a third stage producing 12,000 pounds of thrust, lifted into orbit carrying Yuri Gagarin, the first man in space. From Gagarin's single-orbit flight to Valentina Tereshkova's forty-eight or-

Figure 12. Left: a Russian Vostok (modified R-7), of the type that launched Yuri Gagarin, the first man in space, in 1961. Right: the Mercury-Redstone 3 (MR-3) launching Alan Shepard, the first American in space, in 1961.

bits in 1963, the Soviets sent up five more cosmonauts. In the same year as Tereshkova's mission, Leroy Cooper completed twenty-two orbits in the final U.S. Project Mercury capsule. The Soviet Union thus seemed to be winning the "space race" with its secret booster. But the United States took up the challenge and prepared its own advanced booster. With this vehicle, likewise a product of the building-block technique and a descendant of the V-2, America was determined to do no less than land the first man on the moon.

■ SATURN TO THE MOON

Sputnik was still six months away in April 1957, when von Braun's team initiated design studies for an advanced "Super Jupiter" yielding 1.5 million pounds of thrust. This engine would be able to send payloads of 20,000–40,000 pounds into orbit, or lift 6,000–12,000 pounds to escape velocity. Yet no moon program then existed. Rocketdyne had already designed engines with thrusts of 360,000 pounds (the single-chambered E-1) and even of one million pounds (the F-1), but it would be years before these were built, so von Braun sought off-the-shelf hardware. He and his colleagues decided to cluster eight upgraded Jupiter S-3D engines, to create the Juno 5. The go-ahead was given August 1958, and von Braun renamed the vehicle the Saturn. In September, Rocketdyne was awarded a contract by the Army Ballistic Missile Agency to further upgrade the S-3D, which became known as the H-1. There were to be four H-1 models, the final version producing 205,000 pounds of thrust on lox and RP-1 (kerosene). To save time and money, the original Saturn 1 used a cluster configuration for its fuel reservoir: eight Redstone tanks, each 5.8 feet in diameter, were arranged around a Jupiter tank 8.7 feet in diameter.

During this critical period of the Space Age, there were also a number of major events. NASA was created in 1958; von Braun's team was transferred to NASA in 1960; Saturn became a top priority, with von Braun assuming the directorship of the program; various studies of Saturn's upper stages were undertaken; and in his State of the Union message in May 1961, President John F. Kennedy declared a national spaceflight goal: a manned moon landing before the end of the decade.

The construction of the Saturn's upper stage was an especially troublesome issue. Abe Silverstein, NASA's director of spaceflight development, favored a system using liquid hydrogen and lox, which would yield 40 percent more specific impulse than one using a conventional mixture of lox and kerosene. Silverstein, who had been an associate director of NACA's Lewis Research Center in Cleveland, had been impressed with the center's work on experimental engines (gaseous, as well as some powered by liquid hydrogen and lox) in 1958–1959. Tsiolkovsky, Goddard, and Oberth had earlier calculated the high theoretical performance of lox and hydrogen, but the mixture was still believed to be too volatile to handle easily. Interestingly, within NASA it was von Braun who voiced the greatest opposition to these propellants. His negative attitude extended back to 1937, when Walter Thiel had experienced numerous leaks and handling difficulties with a small lox-hydrogen test motor. The Silverstein Committee, formed to study the problem, finally came up with evidence that convinced von Braun and other opponents to accept lox and hydrogen as the optimum fuel combination for Saturn's upper stages. This proved a major advance in the development of high-energy rocket technology and launch vehicles. The Silverstein Committee had argued that in the long run, conventional fuels would limit the payload capability and the growth potential of launch vehicles. Within a short time (in late 1959), a lox-hydrogen engine manufactured by Pratt & Whitney, the RL-10, was chosen as the upper stage. In standard (non–Saturn V) launch vehicles, an RL-10 upper stage became known as the Centaur.

The Centaur itself had an interesting history. It began as a spin-off of the ultrasecret Project Suntan, an attempt to produce a hydrogen-fueled reconnaissance jet aircraft superior to the U-2. In 1956 Pratt & Whitney undertook to develop this jet; but when Project Suntan was canceled in 1958, the company decided to enter the rocket field. Facing stiff competition from other firms, it sought to offer something different and applied its knowledge of hydrogen to rocket propulsion. At first, the lox-hydrogen engine was meant to enhance ICBM performance. Then Krafft Ehricke, who had been Walter Thiel's assistant at Peenemünde and a spaceflight enthusiast since the 1920s, proposed building a Mars probe using a lox-hydrogen stage atop an Atlas engine. In August 1955 the Air Force awarded a contract for the Atlas-Centaur combination, but the vehicle was intended only for launching advanced reconnaissance and communications satellites. In Saturn I a cluster of six RL-10s, each producing a thrust of 15,000 pounds, powered the second stage, also called the S-IV.

The two-stage Saturn I was a research and development vehicle. It made ten flights, the first in October 1961 with a dummy second stage. Not until the fifth flight, in January 1964, did the first live second stage operate. This stage was capable of sending payloads into orbit, and the remaining five Saturn I's (known as Block II vehicles) orbited test Apollo capsules and Pegasus meteoroid-detecting satellites. The next plateau was the testing of the Saturn IB, which introduced a more powerful second stage: a single lox-hydrogen Rocketdyne J-2, producing 200,000 pounds of thrust.

Since the Saturn was a "national launch vehicle," Pratt and Whitney imparted its knowledge of low-hydrogen systems to Rocketdyne, though there were significant differences between the RL-10 and the J-2. The most important was the J-2's ability to shut down and restart itself in space. Computers played a pivotal role in the design and testing of this complex engine, though Rocketdyne borrowed liberally from its own experience on the Atlas and other early engines for the development of the J-2's turbopumps. The Saturn IB, 225 feet tall, was launched in February 1966. With a new second stage (called the S-IVB), powered by a J-2 engine, the vehicle could place 40,000-pound payloads in orbit. After only three flights, NASA declared that the IB was ready for manned missions. The first was Apollo 7, launched on October 11, 1968: astronauts Walter Schirra, Donn Eisele, and Walter Cunningham made 163 orbits in almost 111 days. (In 1973 the IB far exceeded this in its three Skylab space station missions. And in 1975 a IB was America's vehicle in the famous U.S.-USSR Apollo-Soyuz docking mission.)

The three-stage Saturn V, 363 feet tall, was thus the culmination of years of enormously complicated development. The "V" stood for the first (S-IC) stage's five cavernous F-1 engines, each delivering 1.5 million pounds of thrust, or 7.5 million total. As early as 1955, Rocketdyne had made a feasibility study of the F-1 for the Air Force. In December 1958, in response to Soviet space coups, NASA awarded a contract to Rocketdyne for the engine; it was seen as the country's major hope of catching up, though neither a vehicle nor an application had yet been planned. As a result of its Air Force study, Rocketdyne had a head start developing the engine, and in April 1961 a full-sized prototype was tested. The F-1 had other advantages: it was more reliable, because it followed conventional lines, and it drew on the technology of the H-1, though the difference in size between the two engines was tremendous. The H-1 was 8.8 feet tall, whereas the F-1 was 18.5 feet; the H-1's exit diameter was 4.9 feet, while the F-1's was 12.2 feet; and their dry weights were, respectively, 2,100 pounds and 18,416 pounds.

The F-1's size was itself a great innovation that required new manufacturing and testing approaches. Its regeneratively cooled combustion chamber, for example, was formed by hundreds of tubes brazed (soldered) together in a contoured bundle. Around the bundle were a number of bands and liners. As in other regeneratively cooled engines, the fuel flowed through the tubes, cooling the chamber prior to fuel injection. Because of the chamber's massive size, however, and so that it could withstand far higher pressures and temperatures than those of smaller engines, the nickel-alloy tubes were brazed in a huge furnace rather than by conventional techniques (hand-held torch). Furnace brazing also age-hardened the tubes. (The tube-bundle technique itself was very important, but hardly new. Goddard and Sänger had used cooling coils or jackets around the chambers beginning in the early 1930s, though the chambers themselves were not made of contoured tubes. This type of manufacture originated in 1947 as the "spaghetti

Figure 13. A Saturn V preparing to lift off for the moon.

motor,'' developed by Edward Neu and Edward Francisco at Reaction Motors Incorporated. Soon, it was adopted throughout the industry for small and large engines, and can be found today, for example, in the Space Shuttle's main engine.)

The dramatic, thundering first static test of the Saturn V's first stage took place at the George C. Marshall Spaceflight Center in Huntsville, Alabama, on April 16, 1965. The second (S-II) stage, powered by five J-2s, generated a total thrust of one million pounds. The third (S-IVB) stage was a single J-2 that yielded 200,000 pounds of thrust (later upgraded to 230,000 pounds). This awesome amount of power could place a 285,000-pound payload into Earth orbit or send 100,000 pounds to the moon. In addition to its own propulsive units, the Saturn V carried a number of smaller rockets for a variety of tasks. At the very top of the vehicle, above the Apollo capsule, sat the launch escape system tower, with its pitch, tower jettison, and launch escape motors. (The pitch motor controlled the angle between the longitudinal and horizontal reference planes of the launch vehicle.) Apollo's escape system, like that of the old Project Mercury capsule, provided a means of rapidly freeing the manned command module in the event there was a malfunction on the launch pad or early in the flight. For normal flight operations, retro rockets pushed back or slowed down each stage after it had been separated, in order to prevent collisions with the next stage; other motors (called ''ullage'' motors) forced fuel into the bottom of the upper-stage fuel tanks to make starting easier, since fuel floated in low gravity; and attitude control motors regulated the alignment of the third stage. Lunar and return operations called for a number of other rockets, from Lunar Module Ascent and Descent engines to various reaction control rockets for spacecraft positioning.

The Apollo-Saturn program was a magnificent achievement of technical engineering and human ingenuity. After two unmanned evaluation flights (Apollos 4 and 6), the first manned lunar orbital mission was launched on December 21, 1968. This was Apollo 8, carrying astronauts Frank Borman, James Lovell, and William Anders. Apollo 11, launched July 18, 1969, with astronauts Neil Armstrong, Edwin Aldrin, and Michael Collins, made the first manned lunar landing on July 20, 1969. This was followed by six other lunar missions, though Apollo 13 was aborted before it reached the moon, due to a failure of the oxygen storage system. In all, Saturn V enabled a dozen astronauts to explore the surface of Earth's nearest neighbor. President Kennedy's national goal had been fulfilled.

5

MODERN ROCKETS

I n the late 1970s, as the reusable Shuttle geared up to become almost the only American launch vehicle, the word "expendable" entered the U.S. space lexicon. "Expendable" meant one-shot-only launch vehicles, and although this application of the word was new, such rockets had actually been around for a long time. For convenience' sake, the term "expendable" is used here to refer to all non-Shuttle vehicles.

■ THE U.S. EXPENDABLES

Since the earliest years of the Space Age, expendables in the United States have undergone many modifications, though they have maintained their original family names: Atlas, Thor (also called Delta), and Titan. (The Jupiter line served well in the Explorer and early Mercury programs but became extinct in 1961, when it was supplanted by the more powerful Atlas.) The longest-reigning and most versatile of the launch families is the Delta.

The Delta's history began in late 1958, when NASA was in its infancy. The vehicle was then a modified U.S. Air Force Thor IRBM that could use any one of three available types of upper stages: the Able, the Able-Star, and the Agena. But mission requirements rapidly became more stringent. Since neither the Atlas nor the Centaur high-energy upper stage was operational, it was decided to build yet another type of upper stage, though at first this was to be only an interim arrangement. Milton Rosen, then of NASA, referred to the fourth modification by the Greek letter Delta, and the name was adopted officially in January 1959, when NASA signed a contract for the manufacture of more of the stages. In April 1959, when NASA signed a contract with Douglas Aircraft for a dozen Thors with the Delta configuration (this was NASA's first launch vehicle contract), the name Thor-Delta became commonly accepted. Eventually it was shortened simply to Delta. The original Delta was identical to the Thor-Able 1: it had a standard Thor first stage, a revamped Vanguard second stage yielding 7,800 pounds of thrust, and a modified Vanguard solid-fuel third stage motor yielding 2,800 pounds. Its chief advantages over its predecessors were better inertial guidance and a system that made active guidance control possible during the longer coasting periods between second burnout and third-stage ignition, so that maximum velocity could be achieved at higher altitudes. This basic Delta stood 91 feet high; its tapered body had a maximum diameter of 8 feet. On its maiden launch on May 13, 1960, its third stage did not ignite, and the Echo satellite it carried was unable to achieve orbit. But its first completed flight, on August 12, set Echo 1 in orbit. Fifteen other successes followed. NASA ordered more of the vehicles and continued to use them because they were reliable and relatively inexpensive. Performance and payload capacity steadily increased with each model.

Between 1960 and 1988, there were a total of thirteen basic models: the standard Delta; Deltas A, B, C, D, E, J, M, and M-6; the 900, 1,000, and 3,000 series; and the Delta 3916/PAM. The original Delta could carry a 100-pound payload into low Earth orbit (altitudes of 100 to 500-plus miles), also called LEO. In 1962 Thor's MB3 engine, producing 150,000 pounds of thrust, was replaced by an upgraded MB3–block II Thor engine yielding 172,000 pounds of thrust. The result was Delta A, which could lift payloads of 150 pounds. This engine remained in service until the Delta 1000 series was introduced in 1972. Meanwhile, new models were obtained by adopting more powerful second- or third-stage engines, lengthening first-stage propellant tanks (vehicles with such tanks were known as "Long-Tank Thors"), and incorporating improved guidance systems. In 1964 the D version, with three strap-on solid-fuel Castor 1 motors, producing 53,000 pounds of thrust each and a total thrust of 325,000 pounds, became the first Thrust-Augmented Delta (TAD), also called the Thrust-Augmented Improved Delta (TAID). Maximum diameter with boosters was 14 feet 8 inches. The TAD could lift 1,000 pounds to an orbit at 500 miles. Model M6 of

1969 was the first Delta with six Castor strap-ons; three of the Castors burned about thirty seconds after lift-off, increasing the overall burning time of the boosters. Payloads could be up to 1,000 pounds. The 900 series, introduced in 1972, had nine Castors strapped to the first stage, an upgraded second stage, and a more precise guidance system. At launch, six of the Castors ignited along with the central first-stage engine, producing a total thrust of 480,470 pounds. When these Castors burned out, about 40 seconds after lift-off, the remaining Castors ignited; they burned for another 40 seconds and then dropped off, as the Delta's sustainer engine continued its burn. Total burning time was 220 seconds. The two-stage Delta 900 was able to place 3,700-pound payloads into LEO, while the Delta 904 could lift 740 pounds into geosynchronous orbit, or GEO, as described in Chapter 3.

The 1,000 series, which also appeared in 1972, had an extended long-tank configuration and a newer engine. This was the so-called Straight Eight version: the core rocket was now a constant eight feet in diameter, instead of being tapered like the standard Thor. The 2000 model of 1974 adopted the powerful Apollo Lunar Module engine (TRW-201) for the second stage, and a new Rocketdyne RS-27 engine for the first stage. Developed from the H-1, the RS-27 produced 204,600 pounds of thrust, or 514,800 pounds with the addition of six Castor 2s. The two-stage Delta 2910 in this series could place 4,400 pounds into LEO; and the three-stage 2914, up to 1,550 pounds into GEO.

The 2000 series was supposed to have been the last major Delta model, but since the Shuttle would not be ready until 1980, NASA requested the development of the 3000 series, which would be designed for carrying heavier payloads. The 3914 appeared in 1975 and featured Castor IV engines producing 83,600 pounds of thrust each. To moderate the acceleration, five Castor IVs were ignited at lift-off for a total launch thrust of 629,200 pounds. The remaining four Castors were ignited after the other units had burned. The Delta 3914 could place 2,050 pounds in GEO. What was thought to have been the final model, the Delta 3916/PAM (Payload Assist Module), using a Thiokol Star 48 motor for its third stage, was able to place 2,292 pounds in geosynchronous orbit.

In May 1984 NASA transferred the Delta to a private company, Transpace Carriers, Incorporated, and in November it seemed that the last Delta had flown: the Shuttle was to take over NASA's medium-to-heavy satellite traffic. But as it turned out, the Shuttle did not end the

Delta's career after all. As a result of the disastrous Challenger explosion in January 1986, almost all U.S. space launches were grounded for about two and a half years. In desperation, NASA's clients were forced to seek alternative vehicles. Delta came out of retirement, and on March 20, 1987, the Delta 182 carried an Indonesian Palapa communications satellite into orbit. This was the Delta's last commercial payload. But about the same time, the Air Force hired McDonnell Douglas (formerly Douglas Aircraft) to make seven upgraded Delta II vehicles, called Blue Deltas (the ''Blue'' indicating Air Force use), for launching its Navstar nuclear detection satellites. The model used in 1989 was the Delta 3920, capable of lifting 7,610 pounds into LEO, or 2,830 pounds into GEO. Delta II consists of two models, the 6920 and the 7920, both of which have lengthened tanks and a modified RS-27 engine generating 237,000 pounds of thrust. The 7920 will have lighter, more powerful boosters and will be able to put 9,830 pounds into LEO, or 3,560 pounds into GEO. Launch is expected in 1990. An Enhanced Delta II, now on the drawing boards, promises even greater performance: it will be able to lift 11,110 pounds into LEO, and 4,010 pounds into GEO. America's most prolific space workhorse can thus look forward to a whole new life.

In its first twenty-eight years of duty, the Delta launched sixty-seven communications satellites, not to mention dozens of weather, scientific, and biological payloads, for the United States, England, Canada, France, Germany, Indonesia, Japan, Italy, the European Space Agency, and NATO. The Thor was also mated with the Able, the Able-Star, the Agena, the Altair, and Burner 1 and 2. The Thor-Able combination, which led to the Delta, was used briefly from 1958 to 1961, sending up not only Pioneer space probes but also the Explorer 6 and the first Tiros weather satellite. Meanwhile, in 1960, the Thor-Able Star was introduced: this was a two-stage vehicle whose first stage, generating 172,000 pounds of thrust, was an upgraded Thor and whose second stage was powered by a modified Vanguard Aerojet engine. In April 1960, on its launch of the TRANSIT I-B navigation satellite, this vehicle demonstrated the first engine restart in space. During its five-year service, the Thor-Able Star also achieved the first double, triple, and sextuple satellite launches.

The Agena, which was used from 1959 to 1987, was also intended as a restartable upper stage. An outgrowth of Bell Aircraft's Hustler engine, made for the B-58 bomber's Powered Disposable Bomb Pod, the

Agena went through several modifications and became America's most frequently used upper stage, with more than 350 launches to its credit. Most of these have been military payloads using the Atlas booster, although in February 1959 the Agena, in combination with the Thor (as the Thor-Agena) launched the Discoverer I. Scientific satellites like the Orbiting Geophysical Observatory have also been carried with a Thor booster. The Agena B, producing 15,000 pounds of thrust, was the first restartable model and made its successful debut as the Thor-Agena B in December 1960 carrying the Discoverer XVIII. Moreover, there was a Thor-Agena D and a Thorad (for "Thor Addition"), also called the Long-Tank and Thrust-Augmented Thor, or the Thorad Agena D. The Thor-Agena combinations ceased in 1972.

The "new" Altair upper stage, which was combined with the Thor in 1965 to create the Thor-Altair, was actually the old Vanguard third-stage solid-propellant motor. Boeing Aircraft's Burner, the only solid-fuel upper stage with full control and guidance capability, was also adapted to Thor and all other Air Force space boosters, in order to make possible the precise placement of satellites for military navigation and other uses. The Burner I used an Altair motor, whereas the Burner II used the main retro motor from the Surveyor lunar probe, producing 10,000 pounds of thrust. The Burner IIA was a two-stage version whose second stage generated 8,800 pounds of thrust.

The Thor was obviously an extremely versatile first-stage carrier rocket, but the Atlas was much more powerful and was the first American IBCM converted into a space launch vehicle. In the beginning, however, Atlas had its share of problems. The first Atlas-Able, launched in November 1959, failed in its attempt to send a Pioneer around the moon (the payload shroud broke away prematurely, eventually spilling out the third stage and the payload). Two subsequent Atlas-Able Pioneer missions were likewise failures. Project SCORE, an Atlas with a small communications package, had been boosted into LEO in December 1958, but the first successful stage combination came in May 1960, when the Atlas-Agena A launched a large, experimental 5,000-pound Midas II surveillance satellite. Atlas-Agenas thereafter served the space program well until 1973. They were sent on some of America's first deep-space missions in the 400–800-pound class, such as the Ranger and Lunar Orbiter moon probes and the Mariner 1–5 series (Venus and Mars flybys). Assuredly, for Americans, the most nerve-racking Atlas-Agena missions were the orbiting of Agena target and docking vehicles in 1966 by Gemini 8–12 astronauts, to demonstrate manned spacecraft rendezvous and docking maneuvers. Other Earth-orbiting Atlas-Agena launches included the Applications Technology Satellite and the Orbiting Astronomical Observatory.

The Centaur, incorporating two RL-10 engines with 30,000 pounds of thrust total, was the first lox-hydrogen, high-energy, restartable upper stage to be used with a standard expendable launch vehicle. Its first successful flight came in November 1963 (an Atlas-Centaur), but it did not become fully operational until three years later. On October 26, 1966, during an Atlas-Centaur flight, an important milestone was reached: a lox-hydrogen engine for the first time made a full-thrust restart in space. The Atlas-Centaur was capable of sending about 2,000 pounds into deep space, and became even more powerful in the years following. The vehicle made its share of significant and exotic flights in planetary exploration, launching Surveyor soft-landing lunar probes, Mariners 6–10 (Mars, Venus, and Mercury flybys, and a Mars orbiter), Pioneers 10–11 (Jupiter and Saturn flybys), and Pioneer Venus 1–2 (Venus orbiters and a Venus lander). It also enabled Pioneer 10 to reach the highest launch velocity ever attained: 32,000 miles per hour. The Atlas-Centaur's Earth-orbiting successes included several large Intelsat communications satellites. The Atlas-Centaur ended its career in 1987 but, like the Delta, it thereafter entered a new phase. In May 1988 the improved Atlas-Centaur II won an important contract as a new Air Force medium-lift vehicle. The Atlas's manufacturer (General Dynamics) also offers the Atlas IIA, the "commercial" Atlas-Centaur, for civilian users. The first launches of the Atlas II are scheduled for 1991, and those of the Atlas IIA for 1990. Both vehicles offer much higher performance than their predecessors and can place payloads of 6,100 to 6,400 pounds in GEO (the Atlas I can place 4,950 to 5,150 pounds). Today's commercial Atlas-Centaur stands 137 feet tall, has a lift-off weight of about 360,000 pounds, develops a lift-off thrust of 470,000 pounds (with the upper Centaur stage producing 33,000 pounds), and can place 5,200 pounds in GEO.

The Titan was the second American ICBM to be converted to a space launcher, and was first used in this capacity for lifting two-man Gemini capsules. A modified two-stage Titan II was chosen for this critical task because of its state-of-the-art nonexplosive yet hypergolic propellants (the fuels were UDMH and hydrazine; the oxidizer was nitrogen tetroxide). As a result of this extra safety factor, Gemini capsules in Project Mercury were fitted with ejector-seat escape systems rather than with escape towers. The Titan was more powerful than the Atlas, with a first-stage thrust of 430,000 pounds and a second-stage thrust of 100,000 pounds. The Gemini-Titan II, 109 feet tall and 10 feet in diameter, successfully launched all ten Gemini spacecraft in 1965–1966.

As Gemini-Titans accrued valuable experience in spaceflights, more powerful Titans were developed. In 1964 the Titan IIIA and IIIC were introduced for lifting heavier payloads into LEO (20,000 pounds to altitudes of 115 miles, or 13,000 pounds to 1,150 miles). In accordance with the building-block concept (an essential element in the development of expendable rockets), the Titan IIIA consisted of a reliable and proven Titan II core linked to a new, restartable 16,000-pound-thrust Transtage (transfer stage) with an integrated control and guidance module. The Titan IIIC had a strikingly different configuration. In addition to the above, it had two massive solid-fuel strap-on boosters manufactured by the United Technology Center; each was 10 feet in diameter and 86 feet long, and generated a thrust of 1.2 million pounds. The boosters were exceptional achievements that drew directly on the technology of the solid-fuel Minuteman and Polaris missile programs, which in turn traced their technological origins back to the Hermes RV-A-10 of the 1950s. The United Technology Center's great innovation, however, was the segmentation of the propellants, devised by the company early in 1959 as a solution to the problem of transporting huge solid-fuel rockets to launch sites. In segmentation, large insulated packages of solid-fuel propellants are individually cast (a total of five in the case of the Titan III boosters), then joined together in the rocket case with hand-placed clevis pins and O-rings. An additional advance was liquid-injection thrust vector control, in which a reactive fluid (nitrogen tetroxide) is injected through forty-eight special valves into the exhaust nozzle to exert a side force or counterforce for steering the rocket. (The original solid-fuel thrust vectoring technique was developed in the late fifties by Aerojet's Werner R. Kirchner for Polaris and Minuteman.)

The first five-segment motor was successfully tested in July 1963, and the first Titan IIIC flight was made in June 1965, the booster segments having been transported on railroad flatcars from California to Cape Canaveral. The boosters were fired first (generating 2.6 million pounds of thrust) and jettisoned; then the liquid-fuel core engine was fired (474,000 pounds of thrust); this was followed by the sustainer and Transtage, carrying a dummy test payload of 29,285 pounds. The Titan IIIC became operational immediately. Inevitably, its enormous success marked the beginning of a long and impressive life of service, and led to the development of a family of other Titan III boosters capable of lifting extremely heavy payloads. Among these were the Titan IIID and the Titan IIIE–Centaur.

In addition to carrying military payloads (mainly reconnaissance satellites), these "big birds" accomplished landmark space missions. In 1975 the 160-foot-tall IIIE launched two Viking landers to search for life on Mars, two West German Helios probes to orbit the sun, and two Voyagers to unravel the mysteries of Jupiter, Saturn, Uranus, and the outer reaches of the solar system.

In a now-familiar trend, brought about by the advent of the Shuttle and its subsequent long-term grounding after the Challenger accident, the Titan too became commercial after a quarter-century of service. The commercial III version, 204 feet tall, is also known as the Titan 34D7. This model can deliver 31,700 pounds into LEO using a 13-foot-diameter payload fairing that accommodates spacecraft designed for the Shuttle or other commercial launch vehicles. For this sizable performance, it uses boosters of seven segments and the Air Force's Interim Upper Stage, which was developed for the Shuttle. But on June 14, 1989, the twenty-story-tall Titan IV blasted off from Cape Canaveral on its maiden flight and orbited a top-priority missile-warning satellite. Able to carry up to 39,000 pounds into LEO, or payloads originally designed to fly aboard the Shuttle, the new Titan IV was now America's biggest unmanned rocket and a rival of the Shuttle.

Clearly the big expendables are here to stay, and promise a challenging and exciting future for the U.S. space program. At the same time, however, there has always been a need for lightweight, reliable launch vehicles for modest payloads. The Scout is a case in point. Dubbed "the poor man's rocket" because of its initial low cost ($500,000), the Scout was conceived in July 1957 at NACA's Langley Field in Virginia (now NASA's Langley Research Center). It was 72 feet long and 3.5 feet in diameter, used solid fuel, and was built with off-the-shelf hardware. Designed to lift 150-pound satellites to an altitude of 300 miles, or 100 pounds to 5,000 miles, it could also perform sounding rocket functions. The first stage was an Aerojet Senior, also called an Algol, which produced 118,000 pounds of thrust. Derived from the Polaris sub-

Figure 14. U.S. launch vehicles. Clockwise from top left: a Thor-Delta; the Mercury-Atlas (MA-6) lifting John Glenn into orbit, 1962; a Scout; and a Titan IIIC.

marine-launched IRBM, the Algol was steered by a jetavator—a movable vane that deflected the exhaust gases. The jetavator, invented by Willy Fiedler (who had worked on the V-1), was similar to the jet vanes of the V-2, except that it was located inside the combustion chamber and was used on large solid-fuel rockets. The Scout's second stage was a Castor (manufactured by the Thiokol Chemical Corporation), an improved Sergeant motor producing 50,000 pounds of thrust. The third stage was an Antares motor (13,600 pounds of thrust), and the fourth stage was an Altair (3,000 pounds).

Since the first complete vehicle launch in July 1960, all Scout stages have been upgraded. Air Force variants called the Blue Scout and the Blue Scout Junior are also known. The Scout's first flights were reentry tests and ionospheric probes, though it later launched Explorers 9, 13, and 16 and more than 100 other satellites in international programs such as Ariel and Miranda (UK), ANS (the Netherlands), Aeros and Azur (West Germany), Eole and FR-1 (France), and San Marcos (Italy). Scouts have been launched at three sites: Wallops Island, Virginia; the San Marco sea-based platform off the east coast of Kenya, Africa; and Vandenberg Air Force Base in California. By 1989 the Scout was capable of carrying 450-pound payloads to an altitude of 300 miles. The production of Scouts ceased in 1982, but several vehicles remain in stock for use until 1990 or 1991. And since NASA plans to introduce a small Explorer satellite series, the Scout might well go into production again and continue to be a valued if smaller member of America's launcher family.

■ FOREIGN EXPENDABLES

For more than thirty years, the Soviets have relied on their faithful R-7 basic booster, combining it with modified upper stages (designed by Kosberg or the Isayev Bureau) to launch a variety of missions. Type A-1, which carried Gagarin, also lifted 3,000-pound Meteor weather satellites to altitudes of up to 440 miles and Elektron scientific satellites (always launched in twos, weighing 1,700 pounds total) into highly elliptical orbits of 4,400 miles perigee (closest point to Earth) by 42,000 miles apogee (furthest point). With a new 66,150-pound-thrust Kosberg third stage, plus a 15,435-pound-thrust Isayev escape (or interplanetary) stage, Type A-2e attempted but failed to fly a Mars probe on its first launch, in October 1960. Subsequently, however, it had a number of successes: Venera (Venus) probes, averaging 2,170 pounds; Zond 1–3 deep-space probes of about 1,500 pounds; Luna craft; elliptically orbited Molniya communications satellites (weights unannounced); and 1,850-pound Prognoz satellites. The Prognoz were designed to measure solar radiation, and were lifted into elliptical orbits of 370 miles (perigree) by 124,275 miles (apogee).

The standard Type A-2 launcher, without an escape stage, was apparently first used in the launching of the Cosmos 22 recoverable reconnaissance satellite (weight undisclosed) in November 1963. But it first came to prominence in October 1964, when the three-man, 12,500-pound Voskhod was lifted into LEO. In April 1967 a modified A-2 placed the 15,000-pound Soyuz 1 manned capsule into a 140-mile-apogee orbit. The A-2 continues to lift Soyuz capsules, other manned spacecraft, and large military satellites into LEO and is thus the oldest and most frequently used launch vehicle in the Soviet inventory.

In 1964, with the launching of the Cosmos 1, a new expendable was introduced for smaller Cosmos and Intercosmos 1–9 missions (300 to 1,000 pounds). This vehicle, 98.4 feet tall and 5.4 feet in diameter, was developed by Mikhail K. Yangel and based upon the SS-4 Sandal medium-range ballistic missile. In the West, it was known as the Type B-1, or Small Cosmos Launcher. The first stage used a four-chambered RD-214 engine that produced 158,760 pounds of thrust and burned a mixture of red fuming nitric acid (RFNA) and kerosene. The second stage, an RD-119, burned lox and UDMH, and yielded a thrust of 24,255 pounds.

Yangel also developed the SS-5 Skean IRBM, which in 1964 was converted to the Type C-1 space booster for carrying medium-sized Cosmos and Intercosmos satellites (up to 3,300 pounds). This vehicle, 103.2 feet long and 8 feet in diameter, is powered by a four-chambered RD-216 engine that burns RFNA and UDMH, producing a thrust of about 390,000 pounds. It has a 66,150-pound-thrust second stage and a restartable third stage.

In July 1965 watchers of the Soviet space program were treated to a new surprise: the Proton 1, a 26,896-pound scientific satellite, was lifted to 390 miles by a powerful new booster. The feat was repeated in November 1965 and July 1966. The true dimensions of the Proton's leviathan launcher remained a mystery for years, and provoked even more interest when the Proton 4 satellite, launched in November 1968, proved to be far heavier than the Proton 1 (37,500 pounds). Not until 1984 did the Soviets lift their characteristic veil of secrecy, with the launches of Vega 1 and 2, designed to gather data on Venus and Halley's comet. In the spirit of Glasnost, they released films and data on their most powerful launcher, which, like the satellite, was called the Proton. When all the information was sifted, it turned out that there were three Proton launcher models, classified in the West as the D, the D-1, and the D-1e. The heights of these Protons are 171.7 feet, 188.6 feet, and 195.2 feet, respectively. Maximum diameter for all three models is about 30 feet. All were designed by Korolev and Yangel and follow the cluster-of-clusters approach: six strap-on RD-253 gimbaled engines are arranged around a central core, each producing 331,650 pounds of thrust, or nearly 2 million pounds total. The single-chambered engine, burning hypergolic nitrogen tetroxide and UDMH, was developed between 1961 and 1965 by the Gas Dynamics Lab and the Experimental Design Bureau (OKB). Western space authorities have continued to speculate about the Proton's other stages and about its core. (For example, in the D version, it was unclear whether or not the core ignited on the ground with the boosters; it is now believed that in later versions the core ignited after take-off.) According to current Soviet data, in the latest model the second stage (presumably the core) consists of four unspecified Kosberg single-chambered engines producing a total thrust of about 540,000 pounds, while the third stage is likewise a single-chambered engine of about 135,000 pounds thrust, with a vernier of about 6,740 pounds thrust. Type D-1e supposedly had an additional Kosberg escape stage generating a thrust of 34,177 pounds, with which it launched Zond 4–8 probes, Mars 2–7, Luna 15–23, Venera 9–12, and the Vegas.

The Proton, which is not based on any military vehicles, has also set in orbit military payloads like the Cosmos 1603, Progress tankers (space station supply ships), and the Ekran, Raduga, and Gorizont communications satellites. More recently, in July 1988, two Protons boldly lofted an identical pair of Project Phobos probes to Mars and its moons. Today, the Proton is becoming a commercial vehicle. It is capable of lifting 44,000 pounds to LEO and 4,800 pounds to GEO. The Space Commerce Corporation was subsequently established in Houston, Texas, to market their launch services in the United States, though this venture has not yet received full approval from the U.S. government because of restrictions on technology transfers. Nonethe-

Figure 15. Advertisement for the Proton.

less, Japan, Australia, and many other nations seek cargo space on the giant Proton, which will doubtless continue to be used for years to come.

Since 1967 the Soviets have also introduced Type F vehicles, which were derived from the SS-9 Scarp ICBM and which have been used as experimental maneuverable orbital weapons. Type F-2s have also been regarded as a new generation of Cosmos satellite launchers. (The Cosmos series, comprising about 2,000 satellites since 1962, have accomplished a broad range of scientific and military missions.) Above the F series are the mysterious Type G vehicles, alleged to be "super boosters" capable of producing nearly 12 million pounds of thrust. Type Gs, designed for manned lunar flight, all evidently failed and were lost in immense launch explosions in 1969, 1971, and 1972.

A smaller Cosmos satellite launcher, which the Soviets have named the Tsiklon (Cyclone), is less well-known to Westerners but has been in operation since 1977. It is a three-stage liquid-fuel rocket, 128.8 feet tall and 9.8 feet in diameter at its widest point. Despite its obscurity, the Soviets claim that it has had a high success rate—61 good launches out of 63 firings—and that it is capable of placing up to 8,800 pounds into low Earth orbits and elliptical orbits. It is known to have set the Meteor 2 satellite in orbit. Recently, the Soviets have offered commercial launch services with the Tsiklon, as well as with the Vertikal, which was earlier modified as a satellite launcher. (The Vertikal was modified from the old RS-5.)

The Soviet space program is a relentlessly ambitious one, and so it is not surprising that the USSR has surged ahead with its development efforts. In May 1987 it introduced the Energia (Energy), a booster five times more powerful than any previous Soviet model and unmatched by anything in the West. First launched with a dummy payload, the Energia can place 220,000 pounds into LEO. Beginning in the early 1990s it will be used, in conjunction with the USSR's Shuttle, for placing in orbit large multi-man space station modules designed for advanced military and scientific operations. (The United States has its own plans for an Advanced Launch System, which will be capable of lifting payloads of 200,000 pounds to LEO, but this may not materialize for another decade.) The Energia is a booster of the post–Saturn V generation, producing 8.8 million pounds of thrust. Designated Type SL-W by U.S. analysts, the Energia is 198 feet tall, weighs 4.4 million pounds, and consists of four giant strap-on lox-kerosene boosters ar-

Figure 16. A Soviet Energia.

ranged around a central core. This core, 26 feet in diameter, contains four lox-hydrogen engines each producing 440,000 pounds of thrust—a total power that is slightly greater than that of the main engine of America's Space Shuttle. Each booster produces nearly 1.8 million pounds of thrust. Core engines are ignited first; twelve seconds later the boosters are lit and maximum thrust is reached. As in the old Type A and Proton vehicles, the boosters burn only briefly and are jettisoned and recovered. A cargo pod, designed to hold payloads of 300,000 pounds, is mounted piggyback on the core between the boosters. (On manned missions, the Soviet Space Shuttle is mounted here.) In addition to placing massive satellites in Earth orbit, the Energia is also capable of delivering 70,560 pounds to the moon and 60,000 pounds to Mars and Venus. It also has the potential for carrying missions to the outer planets, some of which may possibly be manned flights. Without doubt, as cosmonaut Vitaly Sevastianov has said, the Energia "opens up colossal horizons for fundamental research."

France, which has made tremendous contributions in the history of astronautics, embodied in the works of such writers as Cyrano de Bergerac, Jules Verne, and Robert Esnault-Pelterie, has likewise made its share of recent advances in spaceflight technology. In 1962 the country formed its own space agency, the Centre National d'Etudes Spatiales, and two months later the Ministerial Delegation for Armament initiated the development of a launch vehicle. A mere three years later, France became the third nation in space: on November 26, 1965, at Hammaguir Range in Algeria, it launched an 88-pound Vanguard-class A-1 test satellite. The booster used was a Diamant (Diamond), 62 feet long and 4.6 feet in diameter. The Diamant had evolved from the "precious stones" series of research rockets—Agate, Topaze, and Emeraude—though in its choice of propellant it was also a descendant of the Véronique. The Diamant's Emeraude (Emerald) first stage burned nitric acid and turpentine and produced 66,150 pounds of thrust. Its Topaze second stage, which burned solid fuel, yielded 33,075 pounds of thrust. And its Rubis (Ruby) third stage, likewise solid-fueled, generated 11,685 pounds of thrust. France's second satellite was launched by a U.S. Scout in December 1965, but its third, the Diapason, was set in orbit two months later by another Diamant fired at Hammaguir. The last Algeria-based firing was made in February 1967, after which launch operations were transferred to Kourou in French Guiana. This location was ideal since it was near the equator, and satellite launches could thus take advantage of the Earth's own rotational momentum (the Coriolis effect) for greater launch speed.

At the same time that launches were moved to Kourou, Diamant B was developed. It had larger fuel tanks than its predecessor and a more powerful Valois first stage, which burned nitrogen tetroxide and UDMH, produced a thrust of 77,175 pounds, and increased the payload capacity to 254 pounds. Kourou was inaugurated in March 1970 with the first flight of the Diamant B, carrying a West German DIAL (Diamant Allemand, or Diamond German) satellite. French satellites— a Péole and a Tournesol—were set in orbit in 1971. On its final launch in 1973, however, the Diamant B lost its twin Pollux and Castor satellites. In 1972 the Centre National d'Etudes Spatiales began work on another upgraded booster, the Diamant B/P.4, which could lift 440 pounds into a low Earth orbit with an altitude of 186 miles. The thrust of the Valois was increased to 88,200 pounds for this rocket. In addition, it had a new second-stage engine called the Rita 1, which produced 39,690 pounds of thrust and which derived from the upper stage of the French MSBS fleet ballistic missile. (MSBS stands for Missile Stratégique Ballistique Sol, or Ground Strategic Ballistic Missile.) Like America's Polaris and Scout, the Rita 1 had solid-fuel verniers and used fluid (Freon) injection for thrust vector control. The new Diamant was launched in February 1975 carrying a geodetic Starlette satellite. On subsequent missions it carried backup Castor and Pollux satellites, and also an Aura, which ended the program. In subsequent years, France continued her space program by other means, namely by participating in European Space Agency, Soviet, and other cooperative space programs.

Japan's space program has been dynamic, just like its economy. From the world's smallest test rockets, the Pencil series of 1955, Japan progressed so rapidly that by 1970 it had become the fourth nation in space. On February 11 of that year it launched a 51-pound Ohsumi test scientific satellite, using a four-stage solid-fuel Lambda 4S. This success had not come easily: four previous launch attempts had failed. The Lambda was a descendant of sounding rockets designed by members of the University of Tokyo. It was 54.2 feet tall, had a diameter of

2.4 feet, and generated 42,777 pounds of thrust in its first stage. The Nissan Motor Company built the rocket according to specifications provided by the university's Institute of Space and Aeronautical Science. The four-stage Mu 4S (height 77.4 feet; thrust 174,635 pounds) had actually been intended as the first space launcher but had been held up by technical problems. Also built by Nissan in conjunction with the University of Tokyo, it lofted Japan's second satellite, a 138-pound Tansei, to 685 miles in February 1971. This was followed in 1974 by the first of the Mu variants, the Mu 3C, which had a much more refined guidance and control system: its second stage had Freon liquid-injection thrust vectoring. These so-called M vehicles continued launching satellites until 1975, when Japan's National Space Development Agency, formed in 1969, switched to more powerful liquid-fuel N rockets. The N-1 had, as its first stage, a U.S. Thor-Delta (built in Japan under U.S. license); as its second stage, an 11,905-pound-thrust liquid-fuel engine built by Japan's Mitsubishi Industries; and as its third stage, a Thiokol solid-fuel engine producing 8,820 pounds of thrust (also built in Japan with U.S. permission). Among other achievements, the N-1 placed a Kiku satellite in orbit in September 1975. By 1981 it had been replaced by the N-2, which had nine solid-fuel Castor boosters (total thrust: 640,110 pounds) for lifting payloads into GEO. In 1984–1985 the N-2 enabled Japan to launch its first deep-space probes, the Suisei and the Sakigake, to investigate Halley's comet.

Meanwhile, in 1981, development started on a similar vehicle, the H-1, first launched in August 1986. The H-1 incorporated Japan's first lox-hydrogen second stage and the first Japanese-made inertial guidance system. (Previously, Japan had been dependent upon the United States for its guidance systems, and by agreement was not allowed to examine them.) Mitsubishi's restartable second-stage engine, which produced 23,000 pounds of thrust, was an especially important milestone, since it demonstrated the first-rate space propulsion technology that will be incorporated into the H-2, the first liquid-fuel launch vehicle fully designed by Japanese engineers. Resembling an early Delta, in that it will have a pair of strap-on solid-fuel boosters, the H-2 will feature a first-stage lox-hydrogen engine (95,000 pounds of thrust) and a second-stage engine adapted from the H-1, and will be capable of lifting a 4,000-pound satellite into GEO. Its maiden flight is planned for 1992. Its predecessor, the H-1, thus represented a significant advance in Japan's effort to join the front rank of space nations.

China, although it may have developed primitive gunpowder rockets a millennia ago, became involved quite late in modern liquid-fuel rocket technology. The roots of that technology can possibly be traced to the career of one man, Dr. Hsue-shen Tsien. Born in Shanghai in 1909, Tsien was a brilliant engineering student who came to the United States on a scholarship. In 1935 he studied aerodynamics under von Kármán at the California Institute of Technology, and in 1936 joined Frank Malina's GALCIT Rocket Research Project. His primary task was to work with Malina on the theoretical thermodynamics of the liquid-fuel rocket motor. Tsien is not known to have participated actively in GALCIT experiments, but he did maintain a strong interest in the theoretical possibilities of rockets and became a noted authority. In 1945 he was sent to Germany as part of the U.S. Navy's Technical Commission to study German missile technology, and he subsequently became the first Robert H. Goddard Professor of Aerospace Propulsion at Caltech. During the McCarthy era, however, Tsien was accused of being a Communist and was harassed to such an extent that in 1955 he returned to China. Here, he was named to head the Preparatory Committee of the Institute of Mechanics of the Academy of Sciences. For years it was assumed in the West that he became head of the Chinese rocket program at this time, though a closer look at various biographical sketches indicates he was involved largely with pedagogical and organizational matters concerning the Chinese Academy of Sciences and other institutions. He also undertook theoretical research on computers, became an active member of the Chinese Communist Party, and has written many works on dynamics and mechanics, as well as popular articles on spaceflight. There is no indication in these biographical sketches that he was directly connected with rocketry after 1955, though it is possible he undertook theoretical studies and may have been involved with other areas of Chinese astronautics. In sum, Westerners simply do not know what Tsien's exact role has been, if any, in the early Chinese rocket program.

The Chinese also obtained expertise in modern rocket technology in a more direct way—namely, through their contacts with the Soviets, before China's rift with the USSR in 1959. It appears that during a time of deteriorating political relations with the Soviets, during the mid- to late 1950s, the Chinese were given or had access to two obsolete V-2 derivative Soviet vehicles, the SS-1 (also known in the West as the

Scunner) and the SS-2 (Sibling). The more powerful of these rockets, the SS-2, used the upgraded V-2 type lox-alcohol RD-101 engine of about 80,000 pounds thrust. The Chinese may have also acquired a number of later-generation SS-3 (Shyster) missiles, using either a lox-alcohol or lox-kerosene version of the RD-103 engine with about 100,000 pounds of thrust and having a horizontal range of about 600 miles. After the Soviet advisers left, the responsibility for laying the foundation of modern Chinese rocket development fell to a man named Huang Weilu. "Carrier rocket expert Huang Weilu," as he is called, was educated in Britain as an electrical engineer and by the 1980s had been appointed chief engineer of China's Ministry of Space Industry.

On October 27, 1966, an atomic weapon was carried for the first time by a Chinese missile; the vehicle was launched from Lop Nor, Takla Makan Desert, in northwestern China, and ascended to an altitude of 400 miles. Probably this was China's first ballistic missile. Known in the West as the CSS-1, it had a maximum horizontal range of about 1,000 miles and was based on the Soviet SS-3. Also under development beginning in 1961–1962 were the CSS-2, a 1,500-mile-range IRBM, and the CSS-3, a 4,300-mile-range ICBM. Then, on April 24, 1970, a three-stage version of the CSS-2 (some say CSS-3), renamed the Chang Zheng 1 (Long March 1), launched China's first satellite, the China 1. This was a 380-pound polyhedral ball which broadcast the national anthem, "The East Is Red." The Chang Zheng 1 (CZ-1) stood about 93 feet tall and was 7.3 feet in diameter. Its four-chambered first-stage engine burned storable nitrogen tetroxide and UDMH, and generated a thrust of 330,750 pounds. The single-chambered second stage used the same fuel, while the third stage used solid propellants. In March 1971, the CZ-1 placed the China 2 in orbit, and in July 1975 the China 3 was lifted to orbit by the CZ-2, also designated the FB-1 (Feng Bao 1, or Storms 1). The CZ-2, derived from the CSS-4 ICBM and developed at the Shanghai Xinxin Machinery Factory, has a first stage consisting of four gimbaled YF-2 engines, each producing 154,350 pounds of thrust (617,400 pounds total). Its second stage has a single YF-2, plus four verniers, generating a total of 9,920 pounds of thrust. The CZ-2 was in service until 1981, with several variants known, some designed for launching recoverable reconnaissance satellites (this is speculation), but most for launching earth resources survey satellites (to improve land management). The CZ-3 is essentially a CZ-2 having a restartable

C Z - 1
CSL - 1
LONG MARCH - 1
CHINA 1; 2.

Figure 17. Foreign satellite launchers. Clockwise from upper left: Diamant (France), Long March 1 (China), Lambda-4S (Japan).

third-stage engine capable of generating 11,000 pounds of thrust. It is designed to boost geosynchronous satellites weighing up to 6,620 pounds.

After the CZ-3 had completed only one successful flight (in April 1984, to orbit China's first geosynchronous communications satellite), the Chinese surprised everyone by offering the rocket to other countries for commercial satellites. They have also offered variants of the CZ-2, and indeed have already found a customer—namely Sweden, which intends to use it for its Mailstar satellite. Several other countries have signed reservation agreements with China's Great Wall Industry Corporation, established for leasing the CZ for commercial purposes. Even the modified CZ-1, which is highly accurate in lifting payloads of up to 1,980 pounds into LEO, is available. In the meantime, in September 1988 the more advanced of the Long March series, the CZ-4, made its maiden flight and launched China's first Sun-Synchronous Orbit satellite (SSO). The SSO, also called the Fen-gyun (Wind and Cloud 1), is a meteorological satellite that monitors the weather and is thus extremely important for this vast and largely agricultural country. Clearly the Chinese recognize spaceflight as an important element in achieving their national goals: first, catching up technologically with the advanced nations and, second, promoting economic and scientific growth by "capitalizing" on their reliable, cheaply produced boosters.

Great Britain just barely qualified as the sixth nation to orbit a satellite with its own vehicle. In the late 1950s its Blue Streak IRBM, powered by twin Rolls Royce 137,000-pound-thrust engines (developed with Rocketdyne assistance), could have served as a satellite launcher. Also available was the Black Knight, a ballistic and reentry test vehicle for the Blue Streak that was powered by a Bristol-Siddeley Gamma 301 engine producing 16,500 pounds of thrust. The Royal Aircraft Establishment and other organizations had earlier proposed a low-cost satellite launcher that would have combined the Blue Streak and the Black Knight. However, in April 1960 the Blue Streak was canceled as a military weapon, for budgetary reasons. By 1964 the Black Arrow, a new, nonmilitary vehicle, had been developed by the Royal Aircraft Establishment and industrial firms. A mere 43 feet tall and 6.5 feet in diameter, it consisted of a 52,075-pound-thrust Gamma Type 8 (eight-chambered) engine that used hydrogen peroxide and kerosene; a

15,750-pound-thrust Gamma Type 2 engine (same fuel); and a 4,927-pound-thrust solid-fuel Waxwing motor, built by Bristol Aero. In 1969, at the Woomera Rocket Range in Australia, the first Black Arrow was aborted fifty seconds after take-off. The second test, a suborbital flight, was successful. On its third flight the rocket was to have launched an Orba satellite, but stage two "underburned" by thirteen seconds and the vehicle failed to achieve orbital velocity. In July 1971 the Black Arrow was canceled. Yet on its last flight three months later, on October 28, 1971, a Black Arrow successfully lifted the 160-pound Prospero X-3 into orbit and into British aerospace history.

In 1962 the efforts of Great Britain and other European nations to build their own space programs resulted in the founding of the European Launcher Development Organization (ELDO) and the European Space Research Organization. Though the Blue Streak had been canceled, ELDO used this rocket as the first stage of its Europa 1 space booster. The second stage was a French Coralie, which burned nitrogen tetroxide and UDMH to produce a thrust of 59,535 pounds, and the third stage was a West German Astris, which generated 5,250 pounds of thrust using the same fuel. However, all three launches of the 103-foot vehicle in 1967–1970 failed. The Europa 2 fared no better. This model was identical to its predecessor, except that it had a fourth stage consisting of a French SEP P.6, a solid-fuel engine producing 9,260 pounds of thrust for placing satellites into GEO. (SEP stands for Société Européen de Propulsion.) The first launch of the Europa 2 took place in November 1971 at Kourou. All seemed to go well initially, but seventeen miles into the ascent the new inertial guidance system malfunctioned, causing an unnatural tilt. Structural failure followed, and the vehicle exploded. This ended the ill-fated Europa program. ELDO disbanded in 1973 and was replaced by a new organization, the European Space Agency (ESA).

ESA's Ariane launcher sprang from the ashes of the Europa. In 1972 France had proposed an entirely new first stage for a Europa 3; this design was designated the Ariane in 1973. Features of the Ariane's engines can also be traced to several other vehicles: the Véronique and Vesta sounding rockets, and the first stage of the Diamant. In its final form, the Ariane 1 comprised the following: a first stage of four French SEP Viking 5 engines, which burned nitrogen tetroxide and UDMH and which had an aggregate thrust of 540,245 pounds; a second stage consisting of a SEP Viking 4, which produced 154,350 pounds of thrust and burned the same fuel; and a third stage incorporating a lox-hydrogen SEP HM-7, which generated 13,230 pounds of thrust. The HM-7 was Europe's first cryogenic engine and originated from a 1962 study for an HM-4 Diamant A second stage. All told, eleven nations contributed to the design and construction of the vehicle, with France and West Germany accounting for the largest part. The Ariane 1 was 155.4 feet high and 124 feet in diameter, and could lift 3,860 pounds into GEO.

After two postponements, the Ariane 1 was finally launched at Kourou on December 24, 1979, and carried an 850-pound test CAT (Ariane Technological Capsule) into GEO. Unfortunately, in May 1980 the second flight failed when a first-stage engine lost pressure and the other three engines died. But the test program was eventually completed, and in January 1982 the Ariane was declared operational. Arianespace, the world's first space transportation company (founded March 1980), was put in charge of the production, marketing, and launching of Ariane boosters. The vehicle achieved an impressive series of successes and milestones. In June 1983, for example, an Ariane became the first rocket to send two satellites into GEO simultaneously. More than a mere stunt, this launch demonstrated the Ariane's ability to carry double payloads into GEO, an ability which made the vehicle that much more cost effective. The ninth flight, in May 1984, placed in orbit the telecommunications Spacenet-F1 satellite for the GTE Corporation. This marked the beginning of Ariane's purely commercial launches; it was also the first time a private U.S. satellite had been borne aloft on a non-American vehicle. The tenth flight, in August 1984, inaugurated the Ariane 3, which had higher-thrust Viking first- and second-stage engines and two Italian-made (SNIA Viscosia) strap-on solid-fuel boosters producing 150,000 pounds of thrust. (The Ariane 2 was identical to the Ariane 3 except that it lacked the boosters.) The Ariane 3 is capable of lifting 5,000-pound payloads into GEO.

All seemed to augur well for Ariane's future, since the Arianespace Corporation had signed a very lucrative contract with Intelsat (the International Telecommunications Satellite Corporation) for the launch of its satellites, and since the Ariane vehicle had successfully launched Brazilsat and Arabsat communication satellites, as well as the deep-space probe Giotto toward Halley's comet in July 1985. NASA admitted that it faced stiff competition in the now multi-billion-dollar satellite launch business. Yet 1986 proved to be an absolutely devastating year for Western space-faring nations. It began with the Challenger disaster on January 28; this was followed by Delta and Titan explosions. The Ariane seemed on track with the February 20 launch of a dual payload: a French SPOT remote-sensing satellite and a Swedish Viking scientific probe. But on May 30 the Ariane 2 had an aborted flight, and on November 13 the orbiting third stage of the SPOT vehicle exploded, creating dangerous debris that threatened other satellites. Not until September 1987, after an intensive sixteen-month period of investigation and redesign, was the Ariane launched again. It successfully placed in orbit an Aussat and a Eutelsat—an achievement hailed as a crucial step in the effort to expand the Western European space program. Another critical point was the long-awaited Ariane 4 launch in June 1988, lifting into GEO three satellites with a combined mass of 7,744 pounds. Almost twice as powerful as the Ariane 3, the Ariane 4 has two strap-on solid-fuel boosters, burns 50 percent longer in the first stage, and can send payloads of up to 9,260 pounds into GEO. The Ariane 4 (launched June 15, 1988) promises to be a formidable workhorse, and should enable ESA to corner about half of the world's commercial space launches.

Yet Arianespace is already looking forward to the Ariane 5, which is expected to enter service in 1995–1996. Its first stage will be a lox-hydrogen HM-60 (Vulcan) engine that will generate 220,500 pounds of thrust. The vehicle, designed to orbit the manned Hermès mini-Shuttle, will be capable of lifting 16,000 pounds into GEO or 36,000 pounds into LEO. ESA managers are also working to make the Ariane's first stages recoverable by parachute, and to enlarge the launchers of the Ariane 6 and 7 so that they will be able to carry three payloads into GEO simultaneously. It is certain that the Ariane will remain a major contender in the most recent phase of spaceflight history: the competition among commercial satellite launchers.

Despite India's staggering economic and social woes—or perhaps because of them—the Indian government has devoted substantial resources to developing a space program. The Indian Space Research Organisation (ISRO) was founded in 1969. Indians were primarily interested in Earth resources satellites, designed to improve land management; communications satellites, a useful technology for a subcontinent whose people speak fourteen major languages; television

Figure 18. Foreign satellite launchers. Clockwise from upper left: Black Arrow (UK), Ariane 3 (ESA), and SLV-3 (India).

broadcasting satellites, which would help educate a rural, largely illiterate population; weather satellites, to warn of monsoons and floods; and satellites for gathering scientific data. With the aid of the USSR, a long-time ally, India's first and second scientific satellites (the Aryabhata and the Bhaskara) were placed in orbit by Soviet C-1 launchers on April 19, 1975, and June 7, 1979, respectively.

By 1971, ISRO had already established the Sriharikota Launching Range on Sriharikota, a small coastal island about fifty miles north of Madras. Here were developed the Rohini and Menaka sounding rockets, techniques for satellite tracking, and a four-stage solid-fuel SLV-3 (Satellite Launching Vehicle 3). The SLV-3 was almost a Scout-class vehicle, 75.5 feet long and 3.2 feet in diameter. Its first stage had a thrust of 121,940 pounds; the second stage, 51,820 pounds; the third, 17,860 pounds; and the fourth, 4,850 pounds. Interestingly, its reaction control system operated on RFNA and hydrazine in stage 1 and on hydrazine alone in stage 2. On its first flight, in August 1979, a guidance system malfunction prevented orbiting. But on July 18, 1980, India's third satellite, the 88.2-pound Rohini 1, was placed in orbit. Since then, ISRO has launched six more satellites and has used five first-generation launch vehicles; the latter have strap-on boosters and are designated PSLVs (Polar Satellite Launching Vehicles). The latest vehicle is the five-stage, 80,000-pound ASLV (Augmented Satellite Launching Vehicle), capable of lifting a 300-pound satellite into LEO. Though the ASLV suffered a failed launch in July 1988, Indian space officials were not deterred from their long-range plans and two weeks later opened a liquid-fuel rocket plant at Tanuku, in Andhra Pradesh. In the 1990s India plans to introduce liquid-fuel boosters capable of reaching GEO.

Yet India by no means completes the roster of the satellite-launching nations. On September 19, 1988, Israel became the eighth member of this exclusive club when it orbited a 330-pound Ofek (Horizon 1) satellite for collecting data on solar energy and the Earth's magnetic field; Israel also wished to gain experience toward establishing an operational reconnaissance satellite system. The launch vehicle, called the Shavit (Cosmos), was a modified solid-fuel, Jericho 2, medium-range missile with a reported range of 900 miles. Pakistan, Brazil, Australia, and Republic of Korea are also working hard to develop satellite-launching capability. Perhaps this trend proves that spaceflight is indeed a necessary element of modern technological growth and that it offers great potential for a better life for all nations.

6

SPACE SHUTTLES

There was no direct link between Eugen Sänger's Silver Bird and the Shuttle. But Sänger's concept of reusable spacecraft became well established in the literature, especially since he championed the technique up to his death in 1964 and was a leader in the International Astronautical Federation. Undoubtedly, the Silver Bird inspired many similar craft which eventually did evolve into the Shuttle.

■ THE COLLIER'S ERA

Two early postwar designs based on his work, as well as on the winged V-2s, were by von Braun and Dornberger. Von Braun's idea, the more famous, was introduced in the March 22, 1952, issue of *Collier's* magazine as part of a marvelously illustrated series on manned spaceflight. The articles had grown out of the First Annual Symposium on Space Travel, held in October 1951 at the Hayden Planetarium in New York City. Von Braun proposed a giant wheel-shaped space station orbiting at an altitude of 1,075 miles. To build and service it required a three-stage rocket that was 265 feet tall and ran on a mixture of nitric acid and hydrazine. Its first stage, consisting of fifty-one engines, developed 28 million pounds of thrust; its second stage (thirty-four engines), 3.5 million pounds of thrust; and its third stage (five engines), 440,000 pounds of thrust. Even with a five-man crew plus payload, the top stage had ample fuel for a return trip to Earth. This streamlined, aircraft-like "cabinstage" had two pairs of front and rear delta-shaped wings and retractable landing gear. It was capable of reaching an altitude of 1,075 miles. It was this stage that shuttled crew and equipment between Earth and space, to construct and service the station. Von Braun reasoned that such a craft would consume less fuel than expendable vehicles and would lower construction costs. He also suggested, in his book *Das Marsprojekt (The Mars Project,* 1952) and in *Collier's* (April 30, 1954), similar reusable winged vehicles for exploring Mars. They could be sent from Earth's space station to an orbit around the planet and could establish small orbital fueling and supply bases, before finally mounting a full-scale expedition to the surface.

At about the time von Braun was drafting these concepts (1951–1955), Walter Dornberger and Krafft Ehricke were working at Bell Aircraft on the highly classified Bomi (Bomber-Missile) study for the Air Force. Well aware of Sänger's projects, Dornberger traveled to France in April 1952 to visit Sänger, who was then employed at the Arsenal de l'Astronautique, near Paris, working on ramjets and other projects. Dornberger tried to persuade him to join the Bell team, but he preferred to remain in France. Nevertheless, the Bomi owed much to the Antipodal Bomber. The Bomi was a spaceplane that rode piggyback on a vertically launched, five-engined rocket. Booster and spaceplane engines fired together at launch, but subsequently separated. The plane could fly 8,450 miles per hour at a 40 mile altitude. Like its Antipodal ancestor, the Bomi released its bomb over a distant target, made a 180-degree turn, and rocketed toward home. The project was dropped while still in the theoretical stage, however, since little was then known about high-speed aerodynamics and about overcoming reentry heating. But various rocket-propelled research aircraft developed at this time (the X-1, the X-2, the D-558-II, and especially the X-15) contributed immeasurably toward solving these critical problems.

Another famous early Shuttle scheme, begun in 1949 and presented at the American Rocket Society's annual meeting in 1954, was Darrell C. Romick's Three-Stage Satellite Ferry-Rocket Vehicle. Romick, working with two associates from the Goodyear Aircraft Corporation, envisioned an enormous space colony he simply called a "satellite," which orbited at an altitude of 500 miles. This was serviced by the Meteor, a rocket weighing some 18 million pounds and producing 32–38 million pounds of lift-off thrust. The Meteor ferried men and 70,000-pound cargoes to and from the colony. It was a giant three-stage vehicle, each stage with delta-shaped wings, extendable landing gears, and clamshell nose doors that closed when a stage was separated from the subsequent one. Each stage thus became a reusable piloted glider.

Such plans, of course, were far too grandiose to be practical and had no chance of reaching the hardware stage. Meanwhile, the Air Force was developing three projects similar to the Bomi. These were the Brass Bell, a manned orbital military reconnaissance vehicle; the ROBO (Rocket Bomber); and the Hywards, a manned research aircraft that combined elements of both rocket and glider. In October 1957 the three were consolidated into the Air Force's X-20, known as the Dyna-Soar (Dynamic Soaring) vehicle, so called because it combined ballistic missile lift with the soaring and precise flight control of an airplane. For the same reasons, it was also called a boost-glider. In its brief, controversial career, the Dyna-Soar nearly became the very first Space Shuttle.

▪ THE DYNA-SOAR AND ITS OFFSPRING

In November 1958 the Dyna-Soar became a joint project of NASA and the U.S. Air Force. It was a one-man, heat-resistant (radiation-cooled) glider with delta-shaped wings, a rounded ceramic nosecap, and a flat surface. Its wings had a span of 20 feet, and the whole vehicle was 35 feet long. Boosted into orbit by a Titan II or III with a trans-stage (or orbit-transfer stage), the Dyna-Soar later reentered the atmosphere and landed with the aid of gas jets and retro rockets. In principle, the vehicle was designed to test manned maneuverable reentry and to develop the requisite technology, but beyond this the program was ill-defined. Herein lay one of the project's major weaknesses; the other was its great expense. However, some Air Force planners envisaged several possible futures for the Dyna-Soar: as a manned orbiting reconnaissance craft, a space weapons delivery system, and an antisatellite interceptor.

The Air Force began the program with the most exhaustive wind-tunnel tests in the history of flight, involving about thirty wind tunnels. Full-scale mock-ups were built, air-drops from B-52s were scheduled, and manned and unmanned flights around the world were planned. By June 1961, the first class of six Dyna-Soar pilots had started training in simulators, while special X-20A spacesuits were being devised for

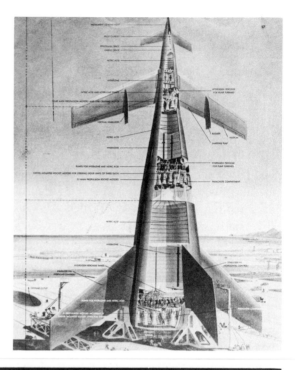

Figure 19. The evolution of Shuttle concepts. Clockwise from lower left: Sänger's Antipodal Bomber, Von Braun's space ferry, Romick's Meteor, and the Dyna-Soar.

them. The Air Force also began airplane flight tests of Dyna-Soar guidance systems and components. From the start, however, the program was continually being reevaluated, especially in light of the upcoming Project Gemini, which could perform similar functions in two-man capsules at less expense. One fact was clear: the Dyna-Soar could not support the future Project Apollo. The delays contributed to the Dyna-Soar's rising cost, which by 1963 nearly equaled that of the entire Mercury program. (The full Dyna-Soar program called for eleven rocket gliders costing $800 million.) The enormous expense and the lack of a program definition inevitably killed the Dyna-Soar, which was canceled in December 1963. Yet the Dyna-Soar contributed a wealth of aerodynamic knowledge and technology toward the development of the Space Shuttle.

A host of other Air Force projects also laid substantial groundwork for the Shuttle. The most important of these were the ASSET and PRIME programs. ASSET (Aerothermodynamic-Elastic Structural Systems Environment Tests) began in 1960 in support of the Dyna-Soar, but was funded separately. The program was designed to test high-speed reentry glide methods and the exotic, heat-resistant refractory metals used in the Dyna-Soar. It consisted of six scaled-down Dyna-Soar winged gliders (minus vertical fins), which plummeted to Earth at 13,000 miles per hour after being boosted to altitudes of 30–50 miles by Thors and Thor-Deltas. The gliders weighed 1,100–1,200 pounds apiece. Built-in transmitters telemetered acceleration, vibration level, pressure, temperature, and other data to a ground station, until the moment each glider landed by drogue parachute. Flight systems included hydrogen peroxide reaction controls and inertial guidance. The first and only ASSET shot for the Dyna-Soar was made in September 1963, since the Dyna-Soar was canceled three months later. But this still did not terminate the program, which was renamed START (Spacecraft Technology Reentry Tests) and which continued highly successfully until 1965.

The PRIME project (Precision Recovery Including Maneuvering Entry) was the second phase of START. Three PRIME vehicles, flown between 1966 and 1967 and boosted into space by Atlas launchers, were likewise small unmanned gliders but were true lifting bodies. (Lifting bodies derive lift from their body shape—usually blunt—and are very useful for testing manned reentry configurations.) PRIME's goal was to demonstrate reentry maneuvering. Its third phase, called

PILOT (Piloted Low-speed Tests), concentrated on refining unpowered pinpoint approaches and landings, at subsonic speeds, from supersonic flights.

In the meantime, from 1963 to 1965, the *manned* M2-F1 lifting body
(initially towed by car or plane) tested landings at very low speeds.
Supersonic manned lifting bodies (the M2-F2, the HL-10, the X-24A,
the M2-F3, and the X-24B) were flown between 1966 and 1975. Usually
air-lifted by B-52s and then released, they were sometimes rocket-
assisted to supersonic speeds before descending and gliding to a landing. The X-24B made especially valuable contributions toward the
Shuttle, not only by gathering data on aerodynamics and handling but
also by test-flying components later incorporated in the Shuttle's orbiter (the manned spaceplane which orbits the Earth). In addition, the
X-24B showed that a spaceplane did not need any cumbersome auxiliary turbojet or other powerplant to land, as some designs of the time
required.

■ THE SHUTTLE TAKES SHAPE

Coincident with the lifting-body period were the glory years of
America's manned space program. Yet NASA was concerned that
after Apollo there would be no viable manned project, whereas the
Soviet Union would surely expand its space activities. Reusable space
transportation, known in principle and studied for years, seemed the
logical answer. Besides this, the spaceplane appeared to offer the private sector and the Department of Defense low-cost launches and
other benefits: satellite maintenance, temporary space laboratories,
and so on. Such were some of the motivations that in April 1969 led
NASA to form the Space Shuttle Task Group, which eventually pur-
suaded President Richard Nixon to approve the Space Shuttle Transportation System (STS) in January 1972. The years 1969 to 1972 thus
represented an intense period of evolution, during which the Shuttle
design was refined on the basis of many proposals.

Several foreign firms also undertook studies of Shuttle-type craft at
this time. The most common configuration was an all-reusable vehicle
composed of both manned booster and orbiter, similar to Romick's
Meteor. For example, Mustard (Multi-Use Space Transport and Recovery Device), developed cooperatively by Germany, France, and

Great Britain, consisted of three piloted spaceplanes joined together. The outside two, serving as boosters, peeled off at altitude and were recovered, while the middle plane orbited, returned to Earth after its mission, and landed on its own.

General Dynamics' Triamese, like the Mustard, consisted of three manned vehicles linked together. The outside boosters had folded wings that deployed on reentry, and extendable turbojets for controlled landings. Among the partly reusable designs was Lockheed's Starclipper, a stage-and-a-half manned lifting body surmounted by two external disposable fuel tanks. On reentry, the remaining lifting body, or orbiter, extended its folding fins for landing. Other designs adapted existing launchers (Titans or Saturn IB's), and still others used alternate propulsion for reusable boosters, such as jets and ramjets.

At first, NASA favored straightforward concepts like North American Rockwell's two-stage, fully reusable Shuttle, which had a large piloted booster and a smaller, piggyback orbiter. It was believed that a very large vehicle was necessary, in order to accommodate the internal propellant tanks. But changing priorities and budgets forced both NASA and industry to revise their plans. Unable to afford a large, fully reusable Shuttle, NASA settled on a partly reusable, smaller vehicle. The breakthrough was NASA's choice of a large external expendable fuel tank. Though expendable, the external tank increased cargo capacity.

The Shuttle, in its working form, has three elements: an orbiter, which is in the center; two large strap-on recoverable solid-fuel boosters (SRBs); and a disposable propellant tank. Each orbiter can be used for about 100 flights, each main engine for about 55 flights, and each SRB for about 20 flights. Launched vertically, the orbiter stands 122.2 feet tall and has a 78.06-foot span across its aircraft-like delta-shaped wings. Its external fuel tank is 154.2 feet long and 27.5 feet in diameter. Running the length of the orbiter is a cargo bay 60 feet long and 15 feet wide, with bay doors for receiving up to 65,000 pounds of cargo, including experimental packages and satellites to be launched. The bay can also be temporarily fitted with a closed or open Spacelab (Space Laboratory). Bay doors are closed at lift-off and open during experimental procedures in orbit. Dwarfing the orbiter and flanking it are two SRBs, each 149 feet long and 12 feet in diameter. SRBs are recovered as follows: after lifting the Shuttle to an altitude of 25 miles, each booster is ejected by small separation rockets, four aft and four forward. Booster descent is slowed by parachutes (a pilot, a drogue, and three main chutes per SRB). After the SRBs land in the water, about 180 miles off Cape Canaveral, the chutes are jettisoned and the SRBs are secured to special recovery ships. The nozzles of the SRBs are plugged to facilitate buoyancy and to protect the rocket-case interiors. The boosters are then towed to port for dismantling, cleaning, and refilling.

After SRB separation, the Space Shuttle's main engines (called SSMEs) continue to burn until the orbiter reaches near-orbital velocity. The external tank then separates and falls into the ocean. Two throttle-controlled, orbital maneuvering system, liquid-fuel engines are fired to complete orbit insertion and later to deorbit; each of these engines produces 6,000 pounds of thrust. Velocity control and positioning movements (pitch, roll, and yaw) are accomplished by forty-four small thrusters fore and aft, collectively called the Reaction Control System.

Upon reentry after a mission, the Shuttle must reduce its speed from 17,600 miles per hour (orbital velocity) to 210 miles per hour for landing. At the fringes of the atmosphere, at an altitude of about 50 miles, the process starts with a series of S-shaped turns to lessen speed and energy. The Shuttle's aerodynamic controls assume responsibility for steering the ship, taking over from the reaction control system. By the time it reaches an altitude of 30 miles, the Shuttle has slowed to Mach 9 (6,750 miles per hour) and begins its third S-turn. At 22 miles the speed is Mach 5 (3,750 miles per hour). At about 15 miles it is Mach 2.5 (1,875 miles per hour). From here, the Shuttle angles its nose downward, glides to about 5.8 miles, and makes a final turn to begin its landing approach. At 400 feet and 210 miles per hour, retractable landing gear is lowered for touchdown on a runway, as with a conventional airplane.

Fully loaded, the Shuttle weighs 4.4 million pounds. Besides cargo, it can lift up to seven people into a typical 115-mile orbit for missions of up to one month. The orbiter's three main engines—throttle-controlled, lox-hydrogen Rocketdyne SSMEs based on technology developed for the Centaur and the J-2—each generate 375,000 pounds of thrust. Each Thiokol SRB produces 2.9 million pounds. Total lift-off thrust is therefore 6.9 million pounds. Confined to low Earth orbits, the Shuttle can reach geosynchronous levels only with the aid of Payload Assist Modules (PAMs) or other extra propulsion stages.

Used for the first time in November 1980, on a Delta launch, the PAM is a reusable cradle assembly with a built-in spin table. The cradle is anchored to the cargo bay keel, and the PAM boost rocket is attached to the satellite; the rocket nozzle is fit into the spin table. When the Shuttle is ready for GEO launch, the spin table is activated. A spring separation system gently pushes the spinning rocket and its attached payload out of the cargo bay. The spinning stabilizes the satellite and prevents it from drifting. After 45 minutes, when the payload and rocket have coasted to a controlled firing point and the Shuttle has moved a safe distance away (20–30 miles), the rocket is fired. This boosts the payload into its desired GEO, where rocket separation occurs.

One remarkable feature of the Shuttle is its reusable thermal protection system. This consists of four types of insulation, which are used on various parts of the ship according to the heat protection required. The most critical areas are the underside and the leading edges of the wings and body, which reach a searing 1,200–2,300 degrees Fahrenheit during reentry. Here are glued 30,922 individually contoured tiles made of silicon carbide and carbon cloth. These tiles, which can be individually replaced if required, are the result of years of research on the part of NASA and the Vought, Lockheed, and Rockwell corporations, beginning in 1959; interestingly, the initial goal of this research was a nonablating, rigid, reusable ceramic insulation for a lifting reentry vehicle. But in the 1970s, this work was adapted to the Shuttle. It was the first time in the history of manned spaceflight that reusable thermal insulation was employed—the Mercury, Gemini, and Apollo craft all had one-use only ablative shields. The Shuttle's insulation system, however, was designed to maintain the integrity of the orbiter through repeated reentries.

■ THE SHUTTLE TAKES FLIGHT

Four Shuttles were authorized: the Columbia, the Challenger, the Discovery, and the Atlantis. The Enterprise, a test orbiter, was introduced in 1977 and initiated crucial free-flight manned gliding trials; it was launched in-flight from the back of a Boeing 747. Then, on April 12, 1981, after numerous postponements due to engine developmental difficulties, John Young and Robert Crippen blasted off from Cape

Figure 20. The modern Space Shuttle. Top: lift-off. Bottom: landing.

Kennedy (formerly Cape Canaveral) in the Columbia, fulfilling Eugen Sänger's fifty-year dream of flying a reusable spacecraft.

After a 54-hour, 37-orbit mission in which the two veteran astronauts tested the STS-1's orbiter systems, including opening and closing the cargo bay doors in space, the ship gently coasted down at Edwards Air Force Base, California, inaugurating a new era of spaceflight. In November, on the second Shuttle test flight (STS-2), the astronauts tested the Canadian-built remote manipulator arm, designed for capturing satellites from orbit and taking them into the cargo bay for repair. The fifth mission, a year later, marked the first fully operational Shuttle flight; with the aid of a PAM, two commercial satellites were successfully launched (the Satellite Business Systems 3 and the Anik C). The first "Getaway Special," a small experimental package, was also carried on this mission. More milestones followed. The first EVA (Extravehicular Activity), or spacewalk, took place on STS-6, which also carried the first six-man crew. On STS-7, in June 1983, Sally Ride became America's first woman astronaut. On the ninth test flight (November–December 1983), John Young commanded the first European Spacelab mission; on board was the first non-U.S. "payload specialist" (scientist), Ulf Merbold of West Germany.

After STS-9, the numbering system of the series was changed. The eleventh flight, for example, was 41C; 4 signified Fiscal Year 1984, 1 stood for a Kennedy Space Center launch, and C indicated the third flight that fiscal year. (Flights following the ill-fated Challenger mission, 51L, reverted back to the original system and picked up with STS-26.)

On 41B, EVAs were performed for the first time on the Manned Maneuvering Unit, a portable astronaut-operated thruster unit on which astronauts moved about without restrictive leash-like cords. On mission 41C, in April 1984, the remote manipulator arm for the first time captured a satellite, the damaged Solar Max, and placed it back into orbit. Flight 61A, in October–November 1985, was the first non-U.S., non-Soviet manned space mission: the D-1 (Deutschland 1) Spacelab, managed by West Germany. This flight also carried the largest crew ever—eight people, including two Germans and a Dutchman.

With the exception of occasional "glitches," such as fallen heatshield tiles on 51D, a failed main engine on 51F, and failure to orbit or repair satellites (41B and 51D), the nation's Space Shuttle Transportation System was running so smoothly that flights almost became routine. The twenty-fifth mission, 51L, changed all that. On January 28, 1986, the Challenger exploded seventy-three seconds after launch, tragically killing its seven crew members. The world grieved at the loss while a special presidential commission, chaired by former secretary of state William P. Rogers, began an exhaustive investigation of the causes of the accident. The commission concluded that the accident had been caused by "failure of a joint between the two lower segments of the right Solid Rocket Motor. The specific failure was the destruction of the seals that are intended to prevent hot gases from leaking through the joint during the propellant burn of the rocket motor." NASA's own investigation identified forty-four other potentially flawed components of the Space Shuttle.

The Challenger disaster halted the U.S. space program for two and a half years, and raised serious questions as to the soundness of its planning. Critics charged that NASA had short-sightedly "placed all its eggs in one basket," by making the Shuttle its sole launcher and phasing out expendables. In August 1986, two months after the Rogers Commission released its report, President Ronald Reagan announced that the Shuttle, when it returned to service in 1988, would carry almost no commercial satellites. The purpose of this decision was threefold: to end backlogged satellite launches, to decrease dependency on the Shuttle by commercial satellite companies, and to open launch services to the private sector—a policy that also aimed to encourage innovation, reduce costs, and promote business growth. Opponents of the decision argued that NASA and the country would lose preeminence in commercial launches, and that private companies would not be able to assume the enormous responsibility of undertaking launches; newer, smaller companies would be particularly susceptible to technical and financial risks. Indeed, when the private sector was opened to highly lucrative Expendable Launch Vehicle (ELV) services, many such companies were founded. But it remains to be seen if they can perform effectively.

Firms competing in this new industry include Space Services, Incorporated, the American Rocket Company (Amroc), Transpace, Pacific American Launch Systems, Incorporated, and the Orbital Sciences Corporation. Several have designed their own rockets, capable of lifting small to moderate payloads. Notable examples are Space Services' solid-fuel Conestoga II; Amroc's Industrial Launch Vehicle, which uses a hybrid system (a series of solid-fuel motors with liquid-fuel injection and ignition); and Orbital Sciences' lightweight Pegasus carrier, which is launched from a B-52 bomber. In fact, Space Services' two-stage solid-fuel Starfire 1 (actually a Terrier-Black Brant sounding rocket), lifted off from the White Sands Missile Range in New Mexico in late March 1989 and entered the history books as the first successful commercial spacecraft launch vehicle. This was a suborbital flight, however, and the 630-pound payload, which contained a series of microgravity experiments for NASA, was retrieved by parachute after a 187-mile-high flight. At the other end of the spectrum, in 1987 the Department of Defense conceived a totally new heavy-lift launch vehicle, designed to free the military of its dependence on the Shuttle. The following year NASA countered with its Shuttle-C (Cargo) design, an unmanned heavy-lift launch vehicle derived from Shuttle hardware; the Shuttle-C uses a cargo container in lieu of an orbiter, but has recoverable SRBs. Both of these launch vehicles have yet to make the transition from mock-up to launch pad.

In fact, in May 1987, as a result of considerable pressure from within and outside the space agency, NASA announced that it, too, would procure ELV services on a competitive commercial basis. This would free the Shuttle for manned scientific missions, national security missions, and functions that only the Shuttle could offer. The reintroduction of expendable launch vehicles would provide NASA with a "mixed fleet" that would support the administration's ELV policy and lessen dependence on a single launch system, namely the Space Shuttle. NASA had thus come full circle, and admitted that expendables were absolutely essential. The agency now saw an immediate need for Deltas, Atlas-Centaurs, and Titans for unmanned scientific payloads, though it was a year before the first contract was awarded under the new policy. According to this contract, the Space Systems Division of General Dynamics would use the Atlas-Centaur to launch a series of Geostationary Operational Environmental Satellites. Meanwhile, the Shuttle was not neglected. With redesigned SRBs and deficiencies corrected, it gained a new lease on life.

Understandably, there was great national anxiety when it came time for the first post-Challenger launch, designated STS-26. Astronauts Frederick H. Hauck, Richard O. Covey, David C. Hilmers, George D. Nelson, and John M. Lounge were selected for this critical mission. The Shuttle Discoverer was to orbit an advanced TDRS-C (Tracking and Data Relay Satellite-C), as well as carry out numerous biological and material-processing experiments. The Discoverer's primary task, however, was to restore confidence in the entire U.S. space program. Failure of the mission was "unthinkable," said one NASA official. To ensure reliability, schedules were not strictly adhered to until everything was absolutely safe for take-off. Vexing delays inevitably occurred; for example, a scratched O-ring seal was discovered in the Inertial Upper Stage, which was to boost the TDRS into GEO, but the scratch turned out to be an improperly smoothed screw mechanism. Another nerve-racking delay occurred thirty-one seconds before take-off, when a faulty computer signaled a loss in cabin oxygen pressure; the problem was quickly solved and the countdown resumed. At 11:37 A.M. EDT on September 29, 1988, thirty-two months after the Challenger disaster, the STS-26 roared America back into space. The four-day mission proved a stunning success, with only ten documented minor malfunctions.

On December 2, 1988, the Atlantis successfully completed the twenty-seventh Shuttle flight: a super-secret military mission that placed in orbit the new Lacrosse surveillance satellite. But space program watchers were well aware that the return of the Shuttle was only a first step toward rebuilding the nation's space program. The Challenger's legacy also meant that the Shuttle could not hope to reach the launch rate for which it was intended. Henceforth, its main task would be to carry out military missions, and it would serve in partnership with expendable vehicles as a member of America's new "mixed fleet."

■ FOREIGN SHUTTLES

The Soviets, like the Americans, early became enamored of Sänger's Antipodal Bomber. In the spring of 1947, Stalin sent his own son Vasily and Colonel Tokaty on a secret mission to kidnap Sänger and his mathematician wife, Irene Sänger-Bredt; but French agents found them first and whisked them from Germany to France, where they settled for a few years. At any rate, it became evident that reusable spacecraft were not yet feasible, though speculative studies of Shuttle-type craft did appear. One example was shown in the animated Soviet film Red Moon (1957), misleading some Western observers to conclude that Sputnik 1 had been launched this way. By the early 1970s, when the United States was formulating its own Shuttle plans, the USSR was engaging in similar studies and appraising different techniques, including ramjet-rocket combinations, flyback boosters, and horizontal take-off designs. By 1978 the Kosmolyot (Spaceplane), as it was now called, had reached the glide test phase. Models of single-seat delta-wing configurations resembling the Dyna-Soar were carried aloft and dropped by Tupolev Tu-95 Bear bombers. (Delta wings derive their name from the fact that a pair of them is shaped like the triangular Greek letter delta.) Rigorous wind-tunnel tests were also conducted at the Central Institute of Aero-Hydrodynamics, outside Moscow. As with the Sputnik and the Proton, however, the Soviets kept this major development under the tightest security.

Western analysts agree that in 1982–1984, the flights of Cosmos vehicles 1374, 1445, 1517, and 1614 were all reentry tests of subscale lifting bodies. Each craft made 1.5 orbits, parachuted into water, and was recovered. The first two ascents were in the southeast Indian Ocean, off Australia; the latter two were in the Black Sea. Royal Australian Air Force photos of the Cosmos 1445's recovery revealed an X-24A-type vehicle with sharply upswept delta wings, a short vertical stabilizer, heat-resistant tiles, a mock cockpit, and evidence of thruster burns. These tests were at first thought to be similar to the ballistic flights of America's PRIME and ASSET programs of 1963–1967. However, this small Soviet vehicle is now believed by some to have been not a scaled-down version of the Soviet Shuttle but a related development, a model of a spaceplane or space ferry. (The latter is a small reusable craft that ferries only a small number of men to a larger orbiting spacecraft like a space station. The space ferry concept is actually an old one that was especially popular in the mid-1950s and 1960s.) Still others think that the little Cosmos 1445 vehicle was indeed a sub-scale Shuttle and that avionics—that is, electronics—and heat-shield material were being tested.

By 1984, it had been reported that full-scale Soviet Shuttles had been built and would soon be ready for approach and landing tests. In 1986, jet engines with a small fuel supply were installed in the Kosmolyot; these would give it greater flexibility in landings than the US. orbiters, which have no means of making a second landing attempt. The Kosmolyot's three lox-hydrogen main engines operated on a disposable external tank, which reduced the orbiter's overall weight, allowed room for the jet engines, and created more payload space. The Soviets later reported that jet engines would not be used on orbital flights, though it now appears that such engines were (and probably still are) used for practice manned landings on a specially fitted Shuttle. For atmospheric landing tests also, the Kosmolyot was initially carried piggyback on a Tu-16 bomber and subsequently on a Bison Mya-4, in a manner similar to NASA's Boeing 747, which carried the U.S. Shuttle on its drop tests. By 1988 the Kosmolyot had made twenty drop tests. In the meantime, the USSR's heavy-lift Energia booster was unveiled, and it seemed likely that this would be the Kosmolyot's operational expendable carrier. This speculation was soon proved correct.

Events progressed rapidly during this period, and on September 29, 1988, the same day the U.S. resumed Shuttle flights with the launch of STS-26, the Soviets at last unveiled their own Shuttle. The USSR's TASS Press Agency released a photo of a vehicle that looked startlingly like the U.S. Shuttle. It was a white delta-shaped craft later called the Buran (Snowstorm), whose underside and leading edges were covered with black ceramic heat-absorbing tiles (38,000 to the U.S. Shuttle's 34,250). But the craft differed from the U.S. Shuttle in a number of important respects. It was attached to the liquid-fuel Energia, and had no large reusable SRBs like its U.S. counterpart. The massive Energia was its sole booster, though the strap-on boosters on the Energia itself are recoverable by parachute. Western observers were baffled by the overall Shuttle-Energia combination, which seemed excessively complex and not cost-effective. (The Soviets themselves were soon openly complaining about the enormous cost of this program.) There was much speculation as to whether the first launch would be manned or not, though the Soviet Shuttle came with a cockpit and was designed to carry six to eight people.

After suffering delays similar to those that plagued the STS-26, the Buran (which was perhaps a different vehicle from the previously shown Soviet Shuttle) made a triumphant lift-off from Baikonur on November 15, 1988. After two low Earth orbits, it returned to home base and landed on a 2.8-mile reinforced concrete runway. The Buran was unmanned on its maiden flight, not only for purposes of safety but also to test its automatic in-flight and landing systems, features that are not found in the American Shuttle.

Yet the Buran is in fact only one of the Shuttle orbiters in the Soviet Vozdushno-Kosmicheskiy Korabl (VKK), or Air-Spacecraft System. There are at least three other vehicles, one of which is named the Ptichka (Birdie). But using the Buran as a basis, one can come up with data for the typical Soviet Shuttle. Its length is 119.4 feet (versus 122 for the U.S. Shuttle); its height (to keel summit) is 54 feet (compared with 56.4 for the American vehicle); and it has a wingspan of 78.4 feet (versus 78 feet for its U.S. counterpart). The payload bays of the U.S. and Soviet vehicles are identical in length (60 feet) and comparable in diameter (15.4 feet USSR, and 14.7 feet U.S.). Empty weights are 62 tons (USSR) to 68 tons (U.S.); loaded weights are, respectively, 106 tons and 120 tons. Payload capacity, performance parameters, mission duration, and designed lifetimes are also identical or match very closely. Obviously, the Soviets learned much from the American program and bypassed huge development costs, though they claim that they had to solve similar aerodynamic and design problems.

Long before the Challenger, however, the Soviets were questioning the efficiency of Shuttles as opposed to relatively cheap ELVs, and they have been cautious in the deployment of their own Shuttle program. They fully intend to keep their stable of ELVs, plan to use the VKK only in "exceptional cases" (two to four flights a year), and value the flexibility of the liquid-fuel Energia, with its automatic shutdown capability (American SRBs cannot be cut off automatically). Moreover, Soviet shuttles are designed to dock with Mir space stations, and they will probably be used for more extensive space operations like constructing solar power stations or assisting interplanetary missions. Undoubtedly, the Soviets are determined to exploit their new Air-Spacecraft System to write new chapters in their remarkable history of cosmonautics.

France's Hermès spaceplane is more modest than the U.S. Space Shuttle and the Buran, yet is a bold step in the development of the European Space Agency's own reusable transportation system. The project began in April 1977, when the Centre National d'Etudes Spatiales asked the firm of Aérospatiale to investigate the feasibility of launching a manned vehicle atop an Ariane 5. The vehicle was not to exceed 14,330 pounds. Typical Hermès missions, with a crew of two to six, could send 10,000 pounds into 250-mile orbits, or 3,300–5,500 pounds into GEO. Average missions would last seven days, but with reduced crew could be extended to thirty days. The Hermès was less than half the size of the Space Shuttle. Its latest configuration has a wingspan of 33.4 feet and a cabin 11 feet wide. It is to be provided with a 4,500-

pound-thrust solid-fuel main engine for re- and de-orbiting, plus liquid-fuel attitude control and docking thrusters. A solid-fuel escape rocket producing 17,980 pounds of thrust is included in the event of an abort at lift-off. Descent is by parachute if a glide-in approach would be impracticable (that is, if the vehicle should come down over water). Since Western Europe lacks experience in manned space rocket flight, data on thermal protection and other systems are gained from the "open literature" (mainly on the U.S. Space Shuttle).

Not until the summer of 1986 did the European Space Agency finally include the Hermès among its multinational programs, though the Centre National d'Etudes Spatiales would have managerial responsibility. The Hermès is still evolving; its first flight is expected by the late 1990s.

Japan, too, is working toward a compact Hermès-type design for its H-2 booster, optimistically called the HOPE (H-2 Orbiting Plane), while Great Britain has proposed a single-stage-to-orbit vehicle named the HOTOL (Horizontal Take-Off and Landing). And the West Germans are planning a two-stage hypersonic-orbiter combination, appropriately dubbed the Sänger. Eugen Sänger himself could hardly have dreamed of so many varied multinational efforts to build upon his concept. His Silver Bird has come a long way.

7

THE FUTURE OF ROCKETRY

"There is nothing new under the sun" is an axiom that aptly applies to rocketry. Nuclear, ion, and photon propulsion, for instance, are all considered "futuristic" yet are found in the earliest spaceflight writings. Such exotic forms of propulsion also began to appear in science fiction, especially after World War I. (Probably the first science fiction story featuring atomic rocket propulsion was by one of Goddard's favorite authors, Garrett Serviss, who wrote *A Columbus of Space,* which initially appeared in serial form in 1909.) This link to science fiction is understandable, since technologically futuristic stories of this genre are usually extrapolations of existing scientific knowledge. The writers were actually somewhat behind the times. As early as the 1890s Antoine Becquerel had discovered radioactivity, and the researches of Marie and Pierre Curie and others had shown that radioactivity was a source of concentrated energy. And between 1900 and World War I appeared the revolutionary theories of Albert Einstein and Max Planck, revealing the nature of both atomic and light (or photon) energy.

■ ATOMIC PROPULSION

Among the early astronautical theorists, Goddard explored the possibility of atomic energy for interplanetary flight in 1906; Tsiolkovsky suggested atomic rockets in 1911; French physician André Bing patented the idea the same year; and Esnault-Pelterie investigated it mathematically in 1912. During the war, around 1942, Ehricke at Peenemünde evaluated German nuclear development for possible application to rocket propulsion. But, he recalled, "I didn't get very far . . . , for work on nuclear physics was much less developed in Germany than it was in the United States." After 1943, Peenemünde's managers contacted Ehricke's former teacher Werner Heisenberg, an eminent nuclear physicist and head of Germany's wartime atomic bomb project, but he could offer no workable proposals.

When the Atomic Age dawned in 1945, countless plans for nuclear-powered spacecraft inevitably emerged, but it was not until November 1955 that the U.S. Atomic Energy Commission (AEC) and the U.S. Air Force initiated Project Rover, with the aim of developing a nuclear rocket for spaceflight. By 1957 a suitably remote test site had been authorized in the Nevada desert, at Jackass Flats, about ninety miles northwest of Las Vegas. This was later designated the Nuclear Rocket Development Station.

In comparison to chemical systems, nuclear rocket propulsion theoretically promised to yield much higher specific impulse and therefore higher exhaust velocities. To put it another way, a rocket's exhaust velocity is proportional to its forward velocity, which means that a rocket with a higher exhaust velocity will accelerate much faster and reach its destination sooner, which is absolutely essential for traveling great distances in space. Exhaust velocities of 30,000–35,000 feet per second are possible with solid-core reactors, whereas a sophisticated chemical system like the Shuttle's can produce only 14,500. Specific impulses would be, respectively, 1,000 and 450. Compared with chemical systems, nuclear systems can heat propellants to higher temperatures, which help account for the greater velocities and impulses. In the basic solid-core nuclear rocket engine, a small nuclear reactor is substituted for the combustion chamber. In place of liquid fuel and oxidizer, only liquid hydrogen is circulated through the reactor, where it reaches extremely high temperatures and spurts out of the exhaust nozzle. (Hydrogen is chosen for its low molecular weight.) The engineering challenges are formidable, however, since thousands of degrees of heat and radiation are produced. But even with regenerative cooling, exotic refractory materials with very high melting points must be used and must be prevented from absorbing neutrons. Project Rover was instituted to gain experience in this new technology and to solve such problems long before flight hardware could be contemplated.

Shortly after NASA was established, an agreement was reached whereby NASA and the AEC would jointly direct the nuclear rocket program. This led to the NASA-AEC Space Nuclear Propulsion Office, which was set up in August 1960 and initiated work on a Nuclear Engine for Rocket Vehicle Application (NERVA). The main goal of this work was the Reactor-in-Flight Tests (RIFT), in which a nuclear-powered third stage would be tried on nothing less than a Saturn rocket. RIFT flights were to begin in 1968. Meanwhile, in July 1959, the first experimental nuclear rocket reactor, the Kiwi A, was successfully ground-tested at Jackass Flats on a static stand mounted on a railroad car. This engine was facetiously called the Kiwi (after the flightless bird of New Zealand) because the stand weighed much more than 5,000 pounds, which was the vehicle's thrust.

The railroad stand continued to ground-test the Kiwi and other systems, so that different reactor designs, fuels, refractory metals, and accessories could be evaluated. The program made substantial progress. By 1965, thrusts of more than 50,000 pounds and specific impulses in excess of 750 seconds had been reached with the Kiwi, the NERVA, and the more advanced Phoebus, which was designed for eventual space travel. In 1967 the NRX-A6 reactor ran for one hour. Reactor flight testing was to follow but never took place. In 1968 budget reductions severely curtailed many of NASA's programs, including its nuclear rocket projects. In 1969 the NERVA-XE raised the specific impulse to 850, but more funding cuts were made, ending all hopes of flight testing. To save the program, NASA proposed that the Space Shuttle carry a NERVA into LEO for a variety of Earth-orbital and deep-space missions. It was a vain attempt. The 1972 budget dealt the final blow. Project Rover was terminated and the station at Jackass Flats was shut down. Thus ended America's active atomic rocket work until 1987, when the Los Alamos Scientific Laboratory in Nevada dusted off the last small Rover engine to begin a new chapter in nuclear propulsion research. If the exploration of deep space is ever to become a reality, say the experts, nuclear propulsion cannot be ignored.

Yet as promising as solid-core systems appear, NASA recognized their limitations and saw them only as a first step. Far higher impulses of 20,000 seconds are possible with gaseous fission engines. Specific impulse increases according to the heat generated by the reactor. In a gaseous-core system, an intensely hot gas fireball is created by fissioning uranium-233 gas or other fuel. Liquid hydrogen is pumped in; it circulates through tubes inside the reactor, changes into superheated gas, and then rushes out the nozzle.

Gaseous-core research began at NASA's Lewis Research Center when work on the NERVA started, and it, too, presented major engineering challenges. One was the problem of how to prevent uranium gas from leaking out of the nozzle. Another was finding a way to provide radiation shielding and neutron reflection, since escaping neutrons cause the reaction to die out. NASA planned to construct miniature gaseous-core engines for working out these problems and then build a full-scale engine. After the demise of Project Rover, gaseous-core studies continued at an experimental reactor at Los Alamos, but by the close of fiscal year 1978 this work, too, had succumbed for lack of funds.

A more controversial fission application was Project Orion, which was established to study the feasibility of a pulse rocket that would work by successive explosions of nuclear bombs. The idea was conceived in 1946 by Polish-born mathematician Stanislaw M. Ulam of the Los Alamos Scientific Laboratory—though Ganswindt's 1890 dynamite spaceship had set the earliest precedent for spaceflight by means of the explosive pulse technique. Ulam had played a key role in the development of the atomic bomb, and he realized that this colossal destructive force could also be harnessed to expand mankind's horizons. His spaceship was designed to be carefully protected and to explode its bombs at a safe distance. The resulting shocks would blast against a pusher plate; in this way the craft would be propelled to the great velocities required for traversing the extreme distances of space.

In 1955, with his coworker Cornelius Everett, Ulam made a landmark mathematical study of the concept and determined that the saucer-shaped craft could reach final velocities of 22,320 miles per hour if thirty one-kiloton bombs were ejected every second and detonated 150 to 1,000 feet behind the ship. Each blast would exert 20,000 pounds of force, the shocks being absorbed through water-cooled springs. The AEC was so intrigued with the concept that it applied for a patent on it. Ulam's idea gained greater acceptance after the launch of Sputnik 1: indeed, in 1958 the Advanced Research Project Agency (ARPA) began funding a study of his spaceship, seeing it as a means by which America might regain its place in space. The ARPA study was afterward transferred to the Air Force and then to NASA, which unofficially and descriptively dubbed it Project Put-Put.

Theodore B. Taylor, head of ARPA's design team, believed that the Orion rocket would be able to transport more than 100 people and many thousands of pounds of cargo to the outer solar system; a Super-Orion, one mile in diameter and propelled by a million bombs, could pulse its way to a nearby star. NASA's version was more modest. It was to be gradually assembled in orbit with the aid of eight two-stage Saturn V rockets. With a specific impulse of 1,850 to 2,550 and expending 2,000 atomic bombs, this 2.5-million-pound Orion was believed capable of carrying twenty people on a 250-day round trip to Mars. Some advocates believed that by 1995 it would be flying fast missions to Venus and Jupiter. The Orion may seem bizarre, but in 1959 and 1960 small flight-test models with high explosives were lifted by small solid-fuel boosters to altitudes of about 150 feet, where they proved the feasibility of pulse rockets. (One model is now in the National Air and Space Museum in Washington, D.C.) The 1963 Nuclear Test Ban Treaty, which outlawed nuclear testing in the atmosphere, appeared to cancel any further possibilities for the Orion. But the true cause of the project's demise, which came in 1965, was budgetary strangulation.

Nuclear fusion, in which light elements are completely fused together, produces much greater amounts of energy than nuclear fission. Estimates of the specific impulse to be derived from fusion range as high as 4.1 million seconds, compared with 1.5 million for fission. Dwain F. Spencer of the Jet Propulsion Laboratory was one of the first to consider fusion for interstellar flight. In 1966 he proposed an engine that would run on deuterium and helium-3, and that would produce extremely hot plasma; the plasma would be directed by superconducting coils toward a nozzle, where it would be forced out, thus propelling the vehicle. Spencer estimated that a ship with such power could reach a star five light-years away in fifty years.

Other proposals soon appeared, the most famous of which was the British Interplanetary Society's Project Daedalus. The Daedalus was designed between 1973 and 1978 by a group of the Society's members headed by Alan Bond, who had worked on Blue Streak rocket engines and on studies of advanced nuclear vehicles. The aim of the project was to devise a practical craft for a one-way unmanned flyby mission to Barnard's Star, six light-years away from the sun. Attention centered on propulsion of the Daedalus. Photon streams, or light rays, were initially considered, but were eliminated as unrealistic because of the staggering powerplant weight and energy that would be required to expel enough light rays for forward thrust. Next, the NERVA's solid-core nuclear engine was examined, but was rejected because of its "limited" exhaust velocity (21,600 miles per hour). An interstellar ramjet, proposed in 1960 by the society's Robert Bussard, was also looked at and discarded; it presented almost insurmountable problems, since it called for collecting interstellar gases to fuel an on-board nuclear reactor. The Daedalus group finally settled on a nuclear pulse engine as the most feasible approach.

In such an engine, controlled fusion pulses would be obtained by ejecting pellets of deuterium and helium-3 into the center of huge (100-foot-diameter) reaction chambers, which would also be magnetic fields. The pellets would be ignited by high-power electron beams, and the resulting 250 explosions per second would propel the ship forward at 2 million miles per hour. This would enable the fully computerized starship to reach its destination in fifty years. Bond's team worked out many of Daedalus' engineering details; for example, the 1.2-billion-pound craft, carrying 1.1 million pounds of automated payload, would have to be assembled in space. They also believed that the ship would be technically feasible by the beginning of the twenty-first century, if one obstacle did not stand in the way: helium-3 is almost nonexistent on Earth, and would have to be obtained by mining the atmosphere of Jupiter. Bond concluded, as have other starship proponents, that interstellar travel will not be realized until the distant future, unless some monumental energy breakthrough is made. Even so, the cost will be astronomical. The venture will have to be undertaken for exploration's sake alone, not for economic reasons.

■ ELECTRIC PROPULSION

In sharp contrast to mammoth starship powerplants are electric propulsion engines, whose thrusts are extremely low but whose specific impulses are high. Impulses of 1,000 and perhaps 10,000 are possible, yet thrust levels are typically smaller than the weight of the rocket by a factor of several thousand. Fuel consumption is minute, but the "push" generated is sufficient in the vacuum and zero-gravity of space. Electric propulsion is therefore well suited for interplanetary voyages, though the engines must be boosted to orbit or escape velocity by a chemical or nuclear vehicle. Electric thrusters also have value as station keepers on satellites and can provide months of continuous service. In electric rockets, the propellant is expelled by electrical energy; there is no combustion, as in conventional rocket engines. The methods of expulsion and the sources of energy are diverse.

In the ion engine, the oldest and most common form of electric rocket engine, electric and magnetic (electromagnetic) forces accelerate atoms or molecules bearing an electric charge (ions) to very high velocities. Goddard suggested using ion rockets for spaceflight as early as 1906. In 1929, Oberth devoted a chapter to them in his *Wege zur Raumschiffahrt*. Also in the 1920s, Ulinski proposed a cathode ray spacecraft in which solar energy was converted to electric power; this power drove a turbocompressor and also fed a cathode ray "ejector," which forced out electrons to propel the ship. Thereafter, little was heard of electric propulsion until 1945, when Herbert Radd of the American Rocket Society suggested using electrically accelerated particles to reduce fuel mass; he appears to have been the first to use the term "ion rocket." And beginning in the 1950s, many other papers on ion propulsion appeared. Ernst Stuhlinger, a member of von Braun's team, emerged as a leading authority and made a number of proposals for manned missions to Mars using electric rockets. Beginning in 1957, perhaps prompted by the success of Sputnik, aerospace companies and agencies also began to investigate experimental ion engines.

In the sixties, one type of ion engine was frequently suggested. In this engine, a stream of atoms from an alkali metal (usually cesium) passed through a hot tungsten electrode, or voltage field. Electrons were pulled away from each atom to form positively charged ions. Several more electrodes accelerated the ion stream. Finally, more electrons were mixed with the ion stream. A neutrally charged beam thus emerged from the engine. (The beam has to be neutral, otherwise the build-up of negative charge in the engine would attract the released positive ions, thereby negating or reducing the thrust of the engine. As with other reaction-propulsion engines, ion engines derive their thrust via reaction to the emission of material from the rear of the engine, in this case a beam of neutral ions. As complex as electric rocket engines can be, they thus all operate according to Newton's classic Third Law of Motion.)

The first successful space electric rocket test (SERT-1) took place on

July 20, 1964. A Scout booster with two ion engines ascended to an altitude of 2,500 miles, verifying the theories and laboratory results. One of the engines was a cesium thruster; the other used a mercury propellant. Although the former suffered a short circuit and did not produce thrust, the latter ran twice and clearly demonstrated that electric engines can perform in space. The 13-pound cesium engine was designed to produce 0.0011 pound of thrust at an exhaust velocity of 176,000 miles per hour; the 11.6-pound mercury engine produced 0.0055 pounds for 30 minutes at an exhaust velocity of 107,000 miles per hour. Both engines were battery operated, but NASA predicted that future electric engines would be powered by either on-board nuclear reactors ("nuclear-electric propulsion") or by solar cells ("solar-electric propulsion," or SEP). SERT-2, which included two SEP thrusters, was launched on February 3, 1970, and demonstrated that solar-electric propulsion was feasible. A nuclear-electric system, however, has still not materialized.

As a result of changing priorities and budget constraints, progress in electric propulsion has been impressive in the laboratory but exceedingly slow in real space. Other types of electric engines have been developed—for example, plasma thrusters. These run on highly ionized gas, or plasma, which is accelerated by electromagnetic fields. Plasma engines were used on the USSR's Yantari 1 geophysical rocket, to an altitude of 250 miles in October 1966, and on the U.S. Navy's Nova 1 satellite, launched in May 1981. Simpler, improved magnetoplasmadynamic (MPD) thrusters, which deliver pulses on the order of milliseconds, were successfully used on Japan's MS-T4 satellite, on Spacelab 1, and on other craft. But although they have had limited uses—such as controlling the USSR's Zond 2 (which failed to reach its destination, Mars)—electric propulsion rockets have not yet served as primary propulsion on any deep-space mission. (NASA's proposed Tempel 2 probe to Halley's comet was designed to be nuclear-propelled, but the project was canceled.)

Still, the future of electric propulsion looks bright, both in the United States and abroad. After conducting its own experimental ion program from 1967 to 1978, Great Britain renewed this work in 1985. It plans to integrate an ion engine with the European Space Agency's Comet Nucleus Sample Return Mission, which may be sent to Comet Wild 2 to gather samples of the comet's surface. In the United States, NASA's proposed multiphase Project Pathfinder includes an MPD-powered

Figure 21. Artist's rendering of a magnetoplasmadynamic (MPD) ship heading toward Mars (Project Pathfinder).

cargo vehicle, designed to support a manned mission to Mars. Toward this ambitious goal, NASA will attempt to develop MPD systems that can generate thousands of watts of power. Electric propulsion may yet prove to be an essential element of long-range interplanetary travel.

■ SOLAR SAILS, PHOTONS, AND LASERS

Just as a schooner on Earth's seas rides before the wind, the solar spacecraft would use a "sail"—a sheet of metallized polymer—to catch the gentle pressure of solar radiation. Theoretically, this form of spacecraft would consume no propellant and would have an infinite specific impulse. The idea sounds romantic and in fact can be traced to an early French science fiction classic, *Aventures extraordinaries d'un savant russe* (*Extraordinary Adventures of a Russian Scientist*), written in 1889 by Georges Le Faure and Henri de Graffigny. In the novel, the heroes fly to the moon and Venus in a sphere encircled by a "selenium" ring, which reflects the rays of the sun. Similar themes are found in other fictional works. We have seen that Goddard made the earliest serious studies of solar propulsion for spaceflight in 1906, while Tsiolkovsky first referred to the idea in 1924 (although he did not elaborate). Also in 1924 the Soviet astronautical pioneer Fridrikh A. Tsander concentrated on this means of propulsion. But probably the first mathematical study of the concept was made in 1958, by Richard L. Garwin of the American Rocket Society.

Looked at critically, solar sails have a number of drawbacks. They would be too slow to traverse the immense distances of space; they would be ineffective in regions beyond Jupiter, where sunlight intensity is low; and they could not be deployed at altitudes of less than 600 miles, because of air drag. A solar sail must be both very light, very large, and highly reflective. Despite the difficulties, many American and Soviet researchers believe that such sails could be used to power cargo ships among the inner planets of the solar system. The World Space Foundation, a private organization started by engineers from the Jet Propulsion Laboratory, plans to build a small solar sail and to test it in orbit. The project would be funded by popular subscription.

Photons (light waves) have also been proposed in fiction and fact as a means of spaceflight propulsion. Hypothetically, a photon rocket would produce the greatest possible specific impulse by traveling at the speed of light (186,000 miles per second). But in 1953, in a series of pioneering studies on photon propulsion, Eugen Sänger found that power requirements would be so high and thrust levels so low, that a rocket operating by photon ejection would be almost impossible to design. Scientists today agree with this thesis. For example, in 1959 Stuhlinger started with the premise that mass can be totally converted into energy. He then calculated that a photon starship with a launch mass of 478,500 pounds, including 368,500 pounds of fuel, would reach nine-tenths of light velocity at the end of a year but would require 475 million megawatts for the fuel-to-energy conversion. To put this into perspective, one should note that in 1957 the world's entire electrical consumption was 350,000 megawatts. Moreover, even if only half the ship's fuel were converted to energy, the temperature of the photon motor would climb to about 10 trillion degrees Kelvin—which would simply vaporize the ship. Photon propulsion thus remains wishful fantasy.

Lasers (Light Amplification by Stimulated Emission of Radiation), introduced in 1960, direct a concentrated beam of light across great distances with extreme precision. Because lasers require ponderous ground-based power sources of millions or even billions of watts, their use in space technology initially seemed limited to deep-space communications, guidance, tracking, and the like. But in 1962, Robert L. Forward of Hughes Research Laboratories proposed that laser rays could be beamed to propulsion systems in space, even to solar sails, and thus could be used to propel spacecraft. Other scientists explored the concept of laser-propelled starships, though it was generally admitted that such a ship would have limited mobility: it would always have to remain within the laser beam. In the 1970s, Arthur Kantrowitz of the Avco Corporation became one of the first to conduct basic laser-propulsion research. He found that specific impulse depends upon laser intensity, propellant density, laser wavelength, and other factors. The powerful rays could instantly turn water into superheated, high-velocity steam and produce enough energy to lift a one-ton spacecraft into orbit. Propellants with low molecular weights, such as hydrogen-alkali mixtures, could be similarly heated by means of a focusing mirror affixed to the spacecraft or vehicle, the mirror being aimed in such a way as to attract the laser beam from a ground-based laser and redirect the beam onto the propellant for ignition. (The mirror, of course, would be suitably protected in the space environment and the beam projected through a transparent port or window in the spacecraft.) The use of pulsed lasers for propulsion has also been suggested. Beginning in 1975, NASA's Lewis Research Center began its own experiments with vacuum chambers and showed that impulses of 1,000 to 2,000 are possible with hydrogen. Stringent cooling requirements, excessive loss of radiant energy, and massive power needs are but some of the technical difficulties still to be solved. But there is no dearth of laser propulsion ideas. Among the more recent are orbital laser and lunar stations for transmitting propulsive power; nuclear-powered lasers located on space stations; lasers activated by solar energy; the use of atomic hydrogen as a propellant; the use of lasers to ignite fusion engines by smashing isotope pellets; and roundtrip interstellar travel by means of laser-pushed solar sails. Today, hampered by limited funds, NASA's laser program is mainly theoretical, concentrating on solar-energy and space-based systems with huge reflecting mirrors. Such systems are feasible and, with proper planning and funding, promise exciting new propulsion developments for deep-space missions.

■ MISCELLANEA: MASS DRIVERS TO ANTIMATTER

As early as 1950, Arthur C. Clarke put forth a serious proposal for an electromagnetic launching rail—an electromagnetic catapult—for use on the Moon. The proposed 8,760-foot-long rail would require 1,000 megawatts of power to hurl, almost instantly, ores and other lunar material at 6,560 feet per second to a peak altitude of about 1,860 miles above the Moon for recovery by a waiting spacecraft. A quarter of a century later, in 1976, a strikingly similar concept was reanimated by Gerard K. O'Neill, a well-known proponent of space colonies. O'Neill's version of this device, now termed "mass driver" and "electromagnetic slingshot," similarly operates on large electromagnetic fields, to accelerate loads of valuable lunar or asteroid ore into predetermined trajectories toward catcher ships. Obviously, such systems could be of service only after lunar and asteroid mining settlements have been established. Based upon the Magneplane, a system combining electromagnetic vehicles and guideways for high-speed travel on Earth, the mass driver is a six-mile track upon which slides a square tubular frame or "bucket." The bucket holds 22 pounds of moon ore. The track may be lined with thousands of electromagnetic coils, so that it becomes a linear electric motor (that is, a track-length electric motor); or it may have a great many solar cells that convert sunlight into electricity, so that it becomes a levitation track. The bucket, containing a superconducting magnet, is propelled along the track by a traveling magnetic wave down until it reaches lunar escape velocity (5,369 miles per hour). In the last 1.2 miles of the ride, the bucket coasts, then opens its lid and flings the moon ore to the receiving ship. Once a bucket has emptied its load, it coasts for another two miles on the track and decelerates. The track then circles back to the loading area where it is reloaded and the process begins again. Similar to mass drivers are rail guns: these are short rails (about thirty feet) that have only one coil. The buckets are not reusable and are ejected with the payload so that bucket deceleration and return is not required.

Many of O'Neill's concepts have evolved in his Princeton classrooms as mind-stretching physics exercises. This is also the case with the Apollo Lightcraft, developed in 1986–1987 by Leik N. Myrabo and his students at the Rensselaer Polytechnic Institute. The Lightcraft is a reusable, beam-powered, single-stage spaceplane that is designed to lift five people (1,100 pounds) to LEO at a far lower cost than the Shuttle can. The 13,200-pound craft would be capable of reaching orbit in three minutes, and could be anywhere on the globe in half an hour. Inspired by the Apollo Command Module, the Lightcraft uses combined-cycle ("mixed") powerplants, which include four "air-breathing" modes and a pure rocket mode for orbital insertion.

Mode 1 is an External-Radiation-Heated (ERH) thruster, used for vertical take-offs up to Mach 3 and for landings. The ERH thruster uses a high-intensity laser beam to create a cylindrical blast wave that propels the ship to a speed of Mach 3 and an altitude of 20,000 feet. (The Lightcraft must point at a laser relay satellite throughout the powered flight.) Modes 2 and 3 are ramjets or scramjets. Ramjets work only when a vehicle is traveling at supersonic speeds; the air is rammed through an intake tube and is consumed as the fuel (hydrocarbon) burns. Scramjets work on the same principle but use liquid hydrogen, which both cools the engine and turns into gaseous hydrogen, propelling the vehicle at speeds of up to Mach 11. Mode 4 is a Magneto-Hydrodynamic (MHD) fanjet, also known as an electric air-turborocket. Liquid hydrogen is superheated by a laser wave. The plasma (ionized gas) produced is driven through an MHD generator, which extracts electric power and accelerates the partially ionized air heated by the shock wave formed at the bow of the craft. This provides the main thrust, but additional thrust comes from the hydrogen exhaust. Mode 5 occurs at altitude of about 40 miles. The MHD generators are shut off, so that the Lightcraft operates solely by means of laser-heated hydrogen rockets. After the craft has coasted to an altitude of about 113 miles, a rocket inserts it into orbit. For reentry the cycles would be reversed, and the command module would serve as a reentry heat shield.

Myrabo sees the combined-cycle approach as "revolutionary," because it could be vastly less expensive and more efficient than present techniques. However, the Lightcraft study assumes the future existence of multiple laser-relay satellites, and does not take into account the enormous costs that would be involved in developing such satellites.

Despite being termed revolutionary, the Apollo Lightcraft actually contains elements that are extrapolations of earlier or current technology. One is the scramjet, which traces its lineage back to the Air Force's Suntan Project of the 1950s. (The Suntan was a high-performance hydrogen plane.) By the 1960s, NASA engineers had theoretically mated this technology with the ramjet to create the scramjet, a supersonic flow and combustion ramjet. Now the scramjet is to become the powerplant for the much-heralded National Aerospace Plane (NASP), also known as the Orient Express, which was announced in President Ronald Reagan's State of the Union address in 1986. The NASP promises to revolutionize air and space travel. It is a kind of super-Shuttle, a one-stage, reusable, multipassenger vehicle that can theoretically take off from an ordinary airport runway, fly at Mach 6 to altitudes of more than 20 miles, and land at another airport a continent away. With a built-in rocket, it could fly from a runway directly into orbit, and return for an airport landing. Cooling the scramjet poses a major problem; it will also be extremely difficult to reach theoretical performance levels. Yet the Aerojet Tech Systems Company has been hired to build the powerplant, and if all goes well, the NASP could be one "future" we may well see in our lifetime.

The ultimate in propulsion, however, would be a ship that travels at the speed of light. Again, the idea is not new, and one can find scientific studies of this concept that were written in the 1950s. The speed is theoretically achieved from the energy released by the annihilation of matter by antimatter. Antimatter is known to science, but only

Figure 22. The future: the National Aerospace Plane.

infinitesimal amounts have been produced; researchers have used powerful linear accelerators, principally at the European Center for Nuclear Research, near Geneva.

In his study for the Air Force entitled ''Antiproton Annihilation Propulsion'' (1985), Robert Forward concluded that antimatter propulsion would be feasible but highly expensive. Another scientist has noted that the production, cooling, and storage of several hundred pounds of antimatter are the major technical obstacles to the success of the concept. More recently, Air Force–sponsored research has shown that if chemicals containing high-energy, high-density matter are added to standard rocket fuels, specific impulse can be dramatically increased. Propellants enriched by antimatter may thus be in use by the year 2000. If controlled antimatter propulsion can indeed be achieved, technological and scientific evolution will have taken a giant leap forward. The human race will be ready for the stars.

SOURCE NOTES

LIST OF ILLUSTRATIONS

INDEX

SOURCE NOTES

The works cited here are only the main sources used in the preparation of this book; these notes are therefore not a definitive guide to the references used. Many of these books contain extensive bibliographies, and suggestions for further reading are also given below.

Introduction

Among the sources used for the "Introduction," the most valuable is N. A. Rynin, *Interplanetary Flight and Communication,* NASA TT F-640 to TT F-648 (Jerusalem: Israel Program for Scientific Translations, 1970–1971), especially the first two volumes. Others are Marjorie H. Nicolson, *Voyages to the Moon* (New York: Macmillan, 1948); and Faith K. Pizor and T. Allan Comp, eds., *The Man in the Moone and Other Lunar Fantasies* (New York: Praeger, 1971). For a more detailed history of the earliest concepts of reaction-propelled flight, see Jules Duhem, *Histoire des origines du vol à réaction* (Paris: Nouvelles Editions Latines, 1959). The analysis of early theories of rocket motion was derived from A. Mandryka, "Fizycne podstawy odrzutu dziala i ruchu rakiety w ujeciu uczonych xvii i xviii wieku" ("The Nature of the Kick of a Gun and of the Motion of a Rocket as Seen by Scientists of the XVII and XVIII Centuries"), *Kwartalnik Historii Naukii Techniki* (Warsaw) 7, no. 4 (1962), and other works.

1. The Founders of Spaceflight Theory

Books in both Russian and English were used for the discussion of Tsiolkovsky, including Arkady Kosmodemyansky, *Konstantin Tsiolkovsky* [in English] (Moscow: Nauka, 1985; though earlier editions are available). One of the better Russian-language biographies is M. Arlazorov, *Tsiolkovskii* (Moscow: Molodaia Gvardina, 1962). The generally unknown connection between Nikolai Fyodorov and Tsiolkovsky was found in Michael Holquist, "The Philosophical Bases of Soviet Space Exploration," *Key Reporter* (Winter 1985): 1–4. I relied on K. E. Tsiolkovsky, *Works on Rocket Technology*, NASA Technical Translation NASA TT F-243 (Washington, D.C.: NASA, 1965), for information on Tsiolkovsky's technical accomplishments. For example, the quote from Tsiolkovsky's *Free Space* (entry of March 28, 1883) comes from this NASA source (p. 3), though *Free Space* itself has not yet been fully translated into English. (See Kosmodemyansky, *Konstantin Tsiolkovsky*, p. 16, for a slightly different translation of the quote.) I also made extensive use of an anthology of Tsiolkovsky's fictional works: Konstantin Tsiolkovsky, *The Call of the Cosmos*, ed. V. Dutt (Moscow: Foreign Languages Publishing House, n.d.).

Much material on Goddard in this chapter comes from Esther C. Goddard and G. Edward Pendray, eds., *The Papers of Robert H. Goddard*, 3 vols. (New York: McGraw-Hill, 1970), which contains selected diary entries, correspondence, reports, and publications by Goddard, along with patent lists and significant achievements. However, the quotations and other details from Goddard's 1902–1909 notebooks have never before been published. I am grateful to the Robert H. Goddard Library, Clark University, Worcester, Massachusetts, for permission to use these invaluable sources. I also consulted the authorized Goddard biography—Milton Lehman, *This High Man* (New York: Farrar, Straus, 1963)—for general biographical information. Translations, including quotes, from Hermann Oberth, *Die Rakete zu den Planetenräumen (The Rocket into Planetary Space)* (Munich: R. Oldenbourg, 1923), are my own. For those who wish to consult this seminal work themselves, a reproduction was published in 1960 by Reproduktionsdruck von Uni-Verlag, Nürnberg. For the treatment of Oberth's 1929 *Wege zur Raumschiffahrt*, I used the English translation *Ways to Spaceflight*, trans. Agence Tunisienne de Public-Relations (Tunis), NASA TT F-622 (Washington, D.C.: NASA, 1972). A reproduction by Kriterion Verlag (Bucharest, 1974) is available. Biographical information was drawn from Hans Barth, *Hermann Oberth: Leben, Werk, Wirkung (Hermann Oberth: Life, Work, Works)* (Feucht, West Germany: Uni-Verlag, 1985). I was also honored, along with Dr. Martin Harwit, director of the National Air and Space Museum, to be able to interview Professor Oberth. Other biographical details, including quotes, come from Hermann Oberth, "My Contributions to Astronautics," in Frederick C. Durant III, and George S. James, eds., *First Steps toward Space*, Smithsonian Annals of Flight no. 10 (Washington, D.C.: Smithsonian Institution Press, 1974), pp. 129–140. The reader should also consult *Hermann Oberth: Briefwechsel (Hermann Oberth: Correspondence)*, ed. Hans Barth, 2 vols. (Kriterion Verlag: Bucharest, 1984), which is a collection of selected correspondence to and from Oberth between 1922 and 1981.

The translation of the quotation from the preface of Konstantin Tsiolkovsky, *Rakyeta v kosmeetcheskoye prostranstvo (Rockets in Cosmic Space)* (Kaluga:

Gosudarstvennaya tipo-litografiya, 1924), is my own. Ilse Essers, *Max Valier: Ein Vorkampfer der Weltraumfahrt, 1895–1930* (Dusseldorf: VDI-Verlag GmbH, 1968), was very helpful in the discussion of Valier's accomplishments. This work was also translated into English by the Agence Tunisienne de Public Relations (Tunis) as *Max Valier: A Pioneer of Space Travel*, NASA TT F-664 (Washington, D.C.: NASA, 1976). Walter Hohmann's classic *Die Erreichbarkeit der Himmelskörper* (1925), translated as *The Attainability of Heavenly Bodies*, NASA TT F-44 (Washington, D.C.: NASA, 1960) by the U.S. Joint Publications Research Service (New York), may be consulted for further information on Hohmann's contributions. See also Heinz Gartmann, *The Men behind the Space Rockets* (New York: David McKay, 1956), with chapters on Ganswindt, Sänger, Valier, et al. The coverage on Esnault-Pelterie is primarily from Lise Blosset, "Robert Esnault-Pelterie: Space Pioneer," in Durant and James, eds., *First Steps toward Space*, pp. 5–31.

2. The Experimenters

Details of Goddard's experiments, with quotations, are drawn from Esther C. Goddard and G. Edward Pendray, eds., *The Papers of Robert H. Goddard*, 3 vols. (New York: McGraw-Hill, 1970); and also from Robert H. Goddard, *Rocket Development Liquid Fuel Rocket Research, 1929–1941*, Esther C. Goddard and G. Edward Pendray, eds. (New York: Prentice-Hall, 1961). *The Papers* includes (vol. 1, pp. 337–406) a reprint of Goddard's classic *Method of Reaching Extreme Altitudes*. My references for other early experimenters are in Frederick C. Durant III, and George S. James, eds., *First Steps toward Space*, Smithsonian Annals of Flight no. 10 (Washington, D.C.: Smithsonian Institution Press, 1974), and include memoirs by Irene Sänger-Bredt, Malina, and several early Soviet experimenters. Additional memoir papers by these and other pioneers of the 1920s and 1930s were found in R. Cargill Hall, ed., *Essays on the History of Rocketry and Astronautics: Proceedings of the Third through Sixth History Symposium of the International Academy of Astronautics*, NASA CP-2014 (Washington, D.C.: NASA, 1977). Coverage on Valier, von Hoefft, von Pirquet, Potočnik (Noordung), Sänger, et al. was also located in Fritz Sykora, "Pioniere der Raketentechnik aus Österreich" ("Pioneers of Rocket Technology from Austria"), *Blätter für Techniksgeschichte* 22 (1960): 194–204. For further reading on the early rocket societies, see Frank H. Winter, *Prelude to the Space Age: The Rocket Societies, 1926–1940* (Washington, D.C.: Smithsonian Institution, 1983). For an excellent English-language account of 1930s Soviet GIRD groups, the reader may also consult Yaroslav Golovanov, *Sergie Korolev: The Apprenticeship of a Space Pioneer* (Moscow: Mir, 1975). An excellent Russian-language source is V. P. Glushko, *Put' v raketnoi teknike: izobrannye trudy 1924–1946 (Road in Rocket Technology: Portrayal of Labor, 1924–1946)* (Moscow: Mashinostroenie, 1977). The von Braun quotes come from Wernher von Braun, "German Rocketry," in Arthur

C. Clarke, ed., *The Coming of the Space Age* (New York: Meredith Press, 1967), p. 39. Eugen Sänger's *Raketenflugtechnik (Rocket Technology)* (1933) has been translated as *Rocket Flight Engineering,* NASA TT F-223 (Washington, D.C.: NASA, 1965), and was used extensively for the analysis of his contributions. The Valier quote in this section is from Ilse Essers, *Max Valier,* German edition.

3. The V-2 Rocket

Unless otherwise mentioned, all quotes by Dornberger come from Walter Dornberger, *V-2* (New York: Viking, 1954), which is considered the standard political-military account of the V-2 weapon. Dornberger's recollections also appear in other sources, but they are not always consistent. Hence, there are several versions of the order he received in 1930 initiating the German Army's rocket program. The version used here is from Walter Dornberger, "The Lessons of Peenemuende," *Astronautics* 3 (March 1958): 18. Compare this with the version in Walter R. Dornberger, "European Rocketry after World War I," in L. J. Carter, ed., *Realities of Space Travel* (New York: McGraw-Hill, 1957), p. 387. For further reading, the technical side of the V-2 is best covered in J. M. J. Kooy and J. W. H. Uytenbogaart, *Ballistics of the Future* (Haarlem, Netherlands: N. V. de Technische Vitgeverij H. Stam, 1946). See also T. Benecke and A. W. Quick, eds., *History of German Guided Missile Development* (Brunswick, Germany: Verlag E. Appelhaus, 1957). Popular technical histories include Gregory P. Kennedy, *Vengeance Weapon 2: The V-2 Guided Missile* (Washington, D.C.: Smithsonian Institution, 1983); and Ernst Klee and Otto Werk, *The Birth of the Missile: The Secrets of Peenemünde* (New York: Dutton, 1963). The story of von Braun's "team" from Raketenflugplatz to Peenemünde to the United States, and of the Soviet Union's use of V-2 hardware and scientists, is covered in Frederick I. Ordway III, and Mitchell R. Sharpe, *The Rocket Team* (New York: Thomas Y. Crowell, 1979). See also Dieter K. Huzel, *Peenemünde to Canaveral* (Englewood Cliffs, N.J.: Prentice-Hall, 1962). Popular though uncritical von Braun biographies are Eric Bergaust, *Wernher von Braun* (Washington, D.C.: National Space Institute, 1976); and Bernd Ruland, *Werner von Braun: Mein Leben für die Raumfahrt (Werner von Braun: My Life for Spaceflight)* (Offenburg: Burda Verlag, 1969). The quote by one of Robert H. Goddard's wartime associates on Goddard's opinion of the V-2 comes from Milton Lehman, *This High Man* (New York: Farrar, Straus, 1963); unfortunately, Lehman does not identify the individual quoted. For British intelligence investigations of the V-2, see David Irving, *The Mare's Nest* (London: William Kimber, 1964). The roundup of V-2 scientists by the Allies is told in James McGovern, *Crossbow and Overcast* (New York: Morrow, 1964); the Russian side is given in Irmgard Gröttrup and S. Hughes, trs., *Rocket Wife* (London: A.

Deutsche, 1959). The Tokaty quote is from G. A. Tokaty, "Soviet Rocket Technology," *Technology and Culture* 4 (Fall 1963): 524.

4. *Rockets Enter the Space Age*

Information on Goddard and sounding rockets was drawn from Esther C. Goddard and G. Edward Pendray, eds., *The Papers of Robert H. Goddard,* 3 vols. (New York: McGraw-Hill, 1970). Data on Erich Regener's experiments with V-2s come from several sources, among them Ernst Klee and Otto Werk, *The Birth of the Missile: The Secrets of Peenemünde* (New York: Dutton, 1963). Details on Malina's work were found in Malina's memoir papers in Frederick C. Durant III, and George S. James, eds., *First Steps toward Space,* Smithsonian Annals of Flight no. 10 (Washington, D.C.: Smithsonian Institution Press, 1974); and R. Cargill Hall, ed., *Essays on the History of Rocketry and Astronautics: Proceedings of the Third through Sixth History Symposium of the International Academy of Astronautics,* NASA CP-2014 (Washington, D.C.: NASA, 1977). Malina's quotes regarding the WAC-Corporal are from Hall, ed., *Essays,* vol. 2, pp. 340, 352. My discussions of the Aerobee and other research vehicles are largely based on the still classic work on the subject, Homer E. Newell, Jr., *Sounding Rockets* (New York: McGraw-Hill, 1959). More recent books that I consulted are William R. Corliss, *NASA Sounding Rockets, 1958–1968* (Washington, D.C.: NASA, 1971); and Jon R. Busse and M. Leffler, *A Compendium of Aerobee Sounding Rocket Launchings from 1959 through 1963* (Washington, D.C.: NASA, 1966). My discussion of the Viking's development in particular was derived from Milton W. Rosen, *The Viking Rocket Story* (London: Faber and Faber, 1956), and from interviews with Rosen. Details of Soviet V-2 derivative developments, and also the Sputnik 1 launcher, were found in M. V. Keldysh, ed., *Tvorcheskoy nasledni akademika Sergei Pavlovich Koroleva (Creative Legacy of Academician Sergei Pavlovich Korolev)* (Moscow: Nauka, 1980), and in various historical articles in the journal *Aviatsia i Kosmonavtika (Aviation and Cosmonautics).* In addition, I consulted V. P. Glushko, ed., *Kosmonavtika: Entsiklopediia (Space Encyclopedia)* (Moscow: Izdatel'stvo sovietskaia entsiklopediia, 1985), and V. P. Glushko, *Rocket Engines GDL-OKB* (Moscow: Novosti Press Agency, 1975). Note, however, that dates of the development of Soviet rocket engines and vehicles differ in these two works. In such cases, I have chosen the dates and other information cited in *Kosmonavtika* because it is a later and more authoritative source. The Sputnik launcher story from the Soviet side can also be found in numerous biographies of Sergei P. Korolev, though these usually lack technical details. An example is P. T. Atashenkov, *Akademik S. P. Korolev (Academician S. P. Korolev)* (Moscow: Mashinostroenie, 1969). See also the biography of Mikhail Yangel, another Soviet rocket designer: Irina V. Strazheva, *Tiul'pani s kosmodroma*

(Tulips from the Cosmodrome) (Moscow: Molodaia Gvardia, 1978). The account of North American Aviation's construction of V-2 engines, and the quotation from George P. Sutton, come directly from an interview with Sutton. Dieter Huzel, Konrad Dannenberg, and other prominent former V-2 rocket technicians also provided considerable help. The official Vanguard history used was Constance M. Green and Milton Lomask, *Vanguard: A History* (Washington, D.C.: NASA, 1970). A firsthand account of America's decision to proceed with the Explorer 1 launch vehicle after Sputnik is John B. Medaris, *Countdown for Decision* (Princeton, N.J.: Van Nostrand, 1959). One very useful source of information on all U.S. launch vehicles during this critical period is Eugene M. Emme, ed., *The History of Rocket Technology* (Detroit: Wayne State University Press, 1964). See also J. Hartt, *Mighty Thor* (New York: Duell, Sloan & Pearce, 1961), for the Thor's early military history. Valuable insights on technological development of the Atlas were provided by John L. Chapman, *Atlas: The Story of a Missile* (New York: Harper, 1960). The best account on Saturn vehicles is Roger E. Bilstein, *Stages to Saturn* (Washington, D.C.: NASA, 1980). Few histories on foreign launch vehicles exist. D. R. Samson, *Development of the Blue Streak Satellite Launcher* (Oxford: Pergamon, 1963), covers the early history of the ill-fated British vehicle. In the area of propulsion, see John L. Sloop, *Liquid Hydrogen as a Propulsion Fuel, 1945–1959* (Washington, D.C.: NASA, 1978); and John D. Clark, *Ignition!* (New Brunswick, N.J.: Rutgers University Press, 1972).

5. Modern Rockets

David Baker, *The Rocket* (New York: Crown, 1978), an excellent comprehensive history of rocketry with an emphasis on military-political aspects, was used throughout this chapter. Willy Ley, *Rockets, Missiles, and Men in Space* (New York: Viking, 1968 and subsequent editions), remains a classic popular history and was also a valuable tool, as was Wernher von Braun and Frederick Ordway III, *History of Rocketry and Space Travel* (New York: Thomas Y. Crowell, 1966). For a useful directory of the world's rockets up to 1960, see Frederick I. Ordway III, and Ronald C. Wakeford, *International Missile and Spacecraft Guide* (New York: McGraw-Hill, 1960). A later directory consulted was Kenneth Gatland, *Missiles and Rockets* (New York: Macmillan, 1975). In addition, the British Interplanetary Society's magazines that featured studies upon specific vehicles served as excellent references, notably Andrew Wilson's series: "Agena: 1959 to 1979," *Journal of the British Interplanetary Society (JBIS)* 34 (July 1981); "Centaur Reaches Fifty," *Spaceflight* 20 (September–October 1978); and "Scout: NASA's Small Satellite Launcher," *Spaceflight* 21 (November 1979). Foreign vehicles are also well covered. On Soviet boosters, see Philip S. Clark, "The Proton Launch Vehicle," *Spaceflight* 19 (September 1977); idem, "The Skean Programme," *Spaceflight* 20 (August 1978); idem, "The Sapwood Launch

Vehicle," *JBIS* 34 (1981); idem, "The Sapwood Launch Vehicle: Revisited," *JBIS* 35 (1982); and idem, "Soviet Launch Vehicles: An Overview," *JBIS* 35 (February 1982). The writings of Valentin P. Glushko on engine development also proved very useful; see Glushko, *Rocket Engines,* cited above, which covers GDL-OKB liquid-fuel rocket engine developments between 1929 and 1974. Other excellent studies are Charles P. Vick, "The Soviet Super Boosters, Parts 1 and 2," *Spaceflight* 15 (December 1973) and 16 (March 1974); and idem, "The Soviet G-1-e Manned Lunar Landing Programme Booster," *JBIS* 38 (January 1985). Historical details of French rocketry came from B. Gire and J. Schibler, "The French National Space Program, 1950–1975," *JBIS* 40 (1987). Details on Chinese launch vehicles, primarily the FB-1, were found in L. Wilbur and James J. Harford, *China Space Report* (New York: American Institute of Aeronautics and Astronautics, 1980). See also P. S. Clark, "The Chinese Space Programme," *JBIS* 37 (May 1984). The biographical sketches on Hsue-shen Tsien are found in *Chinese Communist Who's Who* (Taipei: Institute of International Relations, Republic of China, 1970), s.v. "Qian Xuesen"; *Who's Who in Communist China* (Hong Kong: Union Research Institute, 1966), s.v. "Ch'ien Hsueh-Shen"; and Wolfgang Bartke, *Who's Who in the People's Republic of China* (Armonk, N.Y.: M. E. Sharpe, 1981), s.v. "Ch'ien Hsueh-shen." Press releases by Arianespace Incorporated and others were also useful; see Isabelle Naddeo-Souriau, *Ariane: Le pari européen* (Paris: Editions Hermé, 1986); and Television Digest, Inc., *Ariane vs. Shuttle: The Competition Heats Up* (Washington, D.C.: Television Digest, Inc., 1985). For more on Indian developments, see Neville Kidger, "India's SLV-3 Launch Vehicle," *Spaceflight* 24 (February 1982); Theo Pirard, "India in Space," *Spaceflight* 17 (March 1975); and Mahan Sundra Rajan, *India in Space* (New Delhi: Minister of Information and Broadcasting, 1976).

6. Space Shuttles

The famous *Collier's* piece featuring a Shuttle is Wernher von Braun, "Crossing the Last Frontier," *Collier's* 129 (22 March 1953): 24–29, 72, 74. Brief but important sources on precursors to the Space Shuttle are Curtis Peebles, "Project Bomi," *Spaceflight* 22 (July–August 1980); and Joel W. Powell and Ed Hengeveld, "Asset and Prime: Gliding Re-Entry Test Vehicles," *Spaceflight* 36 (1983). Among numerous sources on the Shuttle, some of the best are John M. Logsdon, "The Space Shuttle Program: A Policy Failure?" *Science* 232 (30 May 1986), which examines crucial policy decisions leading to the Shuttle program; Melvyn Smith, *Space Shuttle* (Sparkford, England: Haynes, 1985), a comprehensive historical treatment that includes the roles of the X-15 and Lifting Body programs; Jerry Grey's thoughtful book *Enterprise* (New York: Morrow, 1979); Marshall H. Kaplan's popular *Space Shuttle: America's Wings to the Future* (Fallbrook, Calif.: Aero, 1978); and William Stockton and John Noble Wilford, *Space Liner* (New York: Times

Books, 1981). For further analyses of the Challenger accident, see Malcolm McConnell, *Challenger: A Major Malfunction* (Garden City, N.Y.: Doubleday, 1987); and Joseph J. Trento, *Prescription for Disaster* (New York: Crown, 1987). Another source is Presidential Commission on the Space Shuttle Challenger Accident, *Report to the Presidential Commission on the Space Shuttle Challenger Accident* (Washington, D.C., 1986), which is quoted in this chapter. Among the variety of news sources used to present the Soviet Shuttle story are *Aviation Week* and *Spaceflight* articles, and the special 67-page Buran edition of *Orbite* (the bulletin of the Cosmos Club of France), Hors Série no. 5 (February 1989), which includes material on test facilities and key personnel, such as the late V. P. Glushko (died 10 January 1989), the "constructor general" of the Energia-Buran system.

7. *The Future of Rocketry*

Krafft Ehricke's quote on nuclear propulsion studies at Peenemünde is from Shirley Thomas, ed., *Men of Space* (Philadelphia: Chilton, 1960), vol. 1, p. 7. Among the historical treatments of futuristic propulsion concepts that I consulted were Saul J. Adelman and Benjamin Adelman, *Bound for the Stars* (Englewood Cliffs, N.J.: Prentice-Hall, 1981), a very readable survey of interplanetary and interstellar flight with sections on nuclear, laser, solar, light, and other propulsion; Kenneth W. Gatland, "Project Orion: America's Semi-Secret Project of the Fifties to Develop a Nuclear Pulse Rocket," *Spaceflight* 16 (December 1974); Anthony R. Martin and Alan Bond, "Nuclear Pulse Propulsion: A Historical Review of an Advanced Propulsion Concept," *Journal of the British Interplanetary Society* 32 (August 1979); and P. M. Molton, "The Nuclear Rocket," *Spaceflight* 12 (October 1970). Robert L. Forward, *Antiproton Annihilation Propulsion* (Edwards Air Force Base, Calif.: Air Force Rocket Propulsion Laboratory, 1985), was an especially interesting find in relating antimatter propulsion. For further reading, there are numerous papers, articles, and books on futuristic propulsion, but for the "historical side" to such concepts, consult, for example, the collection of articles on Daedalus in Alan Bond et al., *Project Daedalus* (London: British Interplanetary Society, 1971); and Louis Friedman, *Starsailing: Solar Sails and Interstellar Travel* (New York: Wiley, 1988). Finally, for an excellent and well-illustrated account of the history of futuristic space concepts from the British point of view, see Bob Parkinson, *High Road to the Moon: From Imagination to Reality,* paintings by R. A. Smith (London: British Interplanetary Society, 1979). Several exotic means of propulsion are included, such as Smith's 1947 plan for a nuclear-powered Moonship, Arthur C. Clarke's 1950 electromagnetic launcher, and Project Daedalus.

ILLUSTRATIONS

INDEX